QB 500 .H55 2005
Hill, Tom.
Space : what now? the past,
present, and possible
futures of activities in
space

Alpine Campus
Library
Bristol Hall
P.O. Box 774688
Steamboat Springs
CO 80477

SPACE: WHAT NOW?

The Past, Present, and Possible Futures of Activities in Space

By Tom Hill
Foreword by Buzz Aldrin

PublishAmerica
Baltimore

© 2005 by Tom Hill.

All rights reserved. No part of this book may be reproduced, stored in a retrieval system or transmitted in any form or by any means without the prior written permission of the publishers, except by a reviewer who may quote brief passages in a review to be printed in a newspaper, magazine or journal.

First printing

Photo of the author by Olivia Schoeller

The opinions expressed herein are the author's, and do not reflect the opinions or policies of the United States Government, The United States Air Force, The Aerospace Corporation, or The Smithsonian Institution.

ISBN: 1-4137-2808-1
PUBLISHED BY PUBLISHAMERICA, LLLP
www.publishamerica.com
Baltimore

Printed in the United States of America

*To my wife, our son and daughter,
in the idea that, if they choose, my children can
participate in some of the adventures described herein.*

*And to everyone on the DC Metro
who wondered what I was typing…*

TABLE OF CONTENTS

Chapter One
Where We Are 1
  The US Government Space Program 1
    Organization 1
    Budget 3
    Headliner Projects 4
    Ongoing Research 6
    Future Plans 7
  The Russian Government Space Program 8
  Europe 10
  Japan 12
  China 13
  India 14
  Treaty Organizations 15
  Space Prizes 15
  "Pure" Entrepreneurships 18
  Non-Profit-Sponsored Activities 22
    Mars Research Stations 22
    Cosmos 1 24

Chapter Two
How We Got Here 25
  Lineage: 1000 – 1957 25
  A Quick Start: 1957 – 1968 28
  Triumph and Waning Interest: July 1969 – 1976 33
  A Space Truck…to Where? Reality Strikes 1977 – 1986 36
  Regaining Footing, but Standing on What? 1987 – 1995 39
  A Bombshell, and Business/Cooperation is (is not) the Answer
    1996 – 2002 45
  Reality Strikes Again, New Horizons: 2003 54

Chapter Three
The *Columbia* Tragedy 61
  January 16 – February 1, 2003, 08:44:09 EST 62
  08:44:09 EST – 09:05 64
  The Search for Answers 66
  NASA's Response During the Mission/Missed Opportunities .. 72

    Underlying Causes 78
        Management Reforms 79
        Contract Turnover (lack of independent viewpoints) 83
    Recommendations 86
    The Road Back 90

Chapter Four
Space and Popular Culture 95
    Comparing Space to "Everyday" Activity 95
        Differences Between Space Activities and Sports 100
    Modern-day Myths 102
        The Astronaut Myth 103
        The "Man-Rated" Myth 111
        The "One-of-a-Kind" Hardware Myth 113
        The "It Must be Complicated to be in Space" Myth 116
        The "If Space Were Important it Would Pay for Itself" Myth . 119
        The Humans vs. Robots Myth 123
    Arguments for and Against Space Travel 126
        For Crewed Space Travel 126
        Against Crewed Space Travel 128

Chapter Five
Space Activism 131
    Roots 131
    A Society for Every Taste 132
        American Institute of Aeronautics and Astronautics 132
        The Planetary Society 133
        The National Space and Satellite Alliance 134
        The Space Frontier Foundation 135
        The B612 Foundation 136
        The National Space Club 137
        The Mars Society 137
        The Moon Society 139
        The Artemis Society 139
        The Mars Institute 140
        ProSpace 141
        Nuclear Groups 142
        The International Committee Against Mars Sample Return .. 144
    Unity: The Answer? 145

Chapter Six

A Robust, Expanding Space Presence 146
   Multiple Methods to Orbit 146
   Two (and a half?) Reusable Stages to Low Earth Orbit 149
   Stand-by Space Rescue/Orbital Safe Havens 150
   On-Orbit Supplies 154
      The Plan 155
      An Orbiting Supply Depot 155
      The Depot 155
      The Delivery Capsule 157
      The Launch Vehicle 157
      The Payoff 158
      The Cost 159
      A Demonstrator Mission 160
      The Issues 160
   A Utility Spacecraft 161
   Separate Cargo (2 types) and Personnel 165

Chapter Seven

Expansion Outward: Where do we go From Low Earth Orbit? 168
   High Earth Orbit 168
   The Moon 179
   The (Near Earth) Asteroids 192
   Mars 201
   No Destination, Just Regular Access 221
   Beyond 224

Chapter Eight

Future Tense – How it Could Happen 225
   Scenario 1: Another Apollo 225
   Scenario 2: Free Enterprise 231
   Scenario 3: A Hybrid Approach 237

Chapter Nine

Yes, but What Should Happen Now? 245
   Set a Goal: Mars, When Private Suppliers can Lift More
   Than Half of the Payload to Low Earth Orbit 245
   Expand Spaceguard 247
   Build a Low-Cost Utility Vehicle 247
   Open Orbital Supply Depots 248
   Complete the Space Station to Meet International Treaties,
   Then Turn it Over to an Authority 248

Discount the Space Station From Future Flights
Beyond LEO (Unless...) 250
After Station Completion, End Crewed Shuttle Flights 251
Consider Removing the Orbiter and Converting
the Shuttle Stack to Cargo Capability 252
Proof-Test Mars Hardware in Low Earth Orbit and High Earth Orbit,
then on Asteroid and Lunar Missions 254
Go to Mars When the Hardware is Ready and Private Enterprise
can meet the Demand 255

Chapter Ten
The New Space Policy of the United States 257
    The Goals and Timelines 257
    The Budget 258
    The President's Commission on Moon, Mars and Beyond 259
    My Opinion 260

Chapter Eleven
The Technical Stuff 264
    How a Rocket Works 264
    Thrust vs. Efficiency 265
    Orbits 267
    Reusability 270
    Getting to Orbit (Why can't I fly from Florida
    and go over the poles?) 272
    Planetary Launch Windows (Why can't I fly to Mars
    whenever I want?) 273
    Slingshots 274
    Nuclear Power in Space 276

Epilogue 281
Glossary 284
Index 300
References 303

FOREWORD

I am one of twelve people in human history that have walked on the moon. When I stood on the plains of the Sea of Tranquility in 1969, I knew that my time on the surface was fleeting. I had no idea that humankind's 20th century interest in the moon was the same way, with no new footfall upon our nearest neighbor since 1972. I have devoted much of my life since 1969 towards the goal of opening space to more people, thereby making it more affordable and allowing human exploitation of the solar system. Through efforts such as ShareSpace, designed to make use of previously unused seats on board the space shuttles, I worked to open NASA's eyes to the potential of flying paying customers on board shuttles that flew below their seven-passenger design limit. Sci-fi author John Barnes and I have written books involving this concept (*The Return*) as well as other ideas relating to how space tourism can open the final frontier to everyone interested (*Encounter with Tiber*).

Times have changed. The Russians have flown the first paying commercial customers on board their spacecraft, and they're scheduled to fly more in the coming years. The loss of the space shuttle *Columbia* sparked a national debate on what the United States wants to do in space, and how much it is willing to spend on that vision. As detailed in the Columbia Accident Investigation Board report, those two decisions are inexorably linked. President Bush set the national space agenda on the 14th of January 2004 for the moon, Mars, and beyond. As I write this, Congress, NASA, and their supporting contractors are busy debating methods and funding to make the vision a reality.

Starting in the 1990s, my path crossed Tom Hill's. At the time, he was in the Air Force, producing a show for the Air Force Academy planetarium based on some of my work on the Tradewind Theory for going to Mars. I saw part of his work, and he captured much of my vision in moving 3D images for planetarium visitors to see. We met again in 1998, while he was a launch control officer at Vandenberg Air Force Base. My theories about Mars travel had changed (the method, not their importance) as had his, but as usual, in sharing ideas both our work became stronger.

Now, he's written this book, and chosen to throw it into the fray of our current national debate on space. This information is necessary. Whether you're an old space salt looking for some new perspective or a newcomer in interest or level of effort to space, I believe you'll find this book full of useful information related to space travel, why we do it, and what we can get out of it.

Tom and I do not see eye to eye on everything. For example, he proposes putting propellant depots in low Earth orbit and using commercial launch services to supply them. I think that the infrastructure would be much more useful at gravity-balance point between the Earth and the moon, but these are trivialities in the scope of human history. We both strongly agree that humankind must look outward to the planets and eventually the stars, moving beyond the cradle of Earth. Tom is a visionary, and has a gift for turning his visions into engaging text while keeping the critical information intact.

So read this book, then recommend it to a friend. If your friend is stingy, lend them your copy. Tell 'em Buzz says it's important stuff that we've got to do.

Buzz Aldrin
Los Angeles, CA
April 2004

INTRODUCTION: FAILED VISION

When I was in elementary school, I found that friends of mine had hobbies. Most of them were sports related (that is, they rooted for the local teams), but others were interested in cars, art, or drawing cars. This sort of thing seemed like a good idea to me, so I decided to get my own hobby. Library period was a big deal back then, and I spent time looking through books on different topics. Personal interest, blind luck, or divine guidance (depending on how you view the world) pushed me into the science section, and eventually I found a small collection of books about space. I'd found my interest, and it became my passion.

In the mid 1970s, publishing was not what it is today, so the selection of current events books (at the time, current events books would have covered the Apollo moon landings, but I know that I didn't see a book on *Skylab* until 1978) was slim. I don't even remember the specific books I read, although I can imagine the scenes where they showed huge-wheeled space stations with spacecraft flying to them frequently. People (given the timeframe, they were likely all men) worked outside in pressurized suits, building the next phase of some huge craft preparing for a mission to the outer planets. The year was typically stated to be 15-20 years in the future. So, simple addition takes us to 1998 on the far end. Hmm.

The vision that was 15-20 years in the future when I was in elementary school is now not even mentioned in most mainstream texts, although it's enjoying a bit of a renaissance in internet writing since the *Columbia* tragedy and President Bush's response. There is talk of technology development, and the promise of leadership (not funding yet) to a far-off goal, but for now, we have to face a question. Yes, we have more hours in space than we did in the early '80s, but to what end?

My life has been devoted to space. Through years of education and work both in the Air Force and as a civilian, I've helped bring the utility of space to people around the world. In my spare time, I've taken on space advocacy.

This has been done on my own, through public speeches; as a member of international organizations, through managing design/build efforts; and on a local level through board membership and small-time project development. I've also visited the halls of Congress, both on my own and as part of a group, trying to give lawmakers (more precisely, their staff members) my vision of a space-faring people. I want to pass on what I've learned.

In this book, I will pull all the threads of space travel together into one coherent message. I will make every effort do so without bias, though the more I see convinces me that any work can and likely will be accused of bias. If an argument is flimsy, I'll say so. If an opinion makes sense to me in my gut, but I can't put a dollars-and-cents argument together for it, I'll say that too. In the end, I do believe this, however: a future where spaceflight is a high priority is a much brighter future for humankind than a one where we remain planet bound.

To start, I want to take a moment and describe what I'm not trying to do. I am not here to give a litany of problems with the United States (or any other nation's) space program. I credit the space program, in an earlier form, for focusing my interests in life. Nothing that's possible today in space would be that way if it weren't for the work done before, and that needs to be recognized. Arguments that things could have been done better are simply sour grapes. Our challenge lies in taking the situation we are in now and building the best future we possibly can for our children and ourselves.

Since I started this journey, my first into the realm of publishing, I've had its limitations brought to light for me in a very stark way. Writing any book about current events is a difficult errand, especially when those events are changing day-to-day. Arguably, the time since February 1st 2003 has been the most volatile years in the space program since its inception, and I've spent the time since then writing this book. At times, current events changed my view on a topic, and I had to weigh changing my description of that event within these pages. Other times, new events eclipsed things that I'd already written (compare my ideas of what should be done now vs. President Bush's new vision for NASA for an example of this). All of this takes place with a looming publishing deadline and the delays between manuscript submission and actual publication. Luckily, modern technology has an answer for this. As the stories covered in this book change, I'll be chronicling them on a website named http://www.spacewhatnow.com. I hope to hear from you there.

For now, let's start off by realizing where we are....

Chapter One

Where We Are

As I write this book in 2003 and 2004, the human space effort is in disarray. The International Space Station, arguably the most important, but unarguably the largest ongoing space effort, is in a caretaker status awaiting a space shuttle return-to flight. A promising speech made by President Bush is facing mixed reactions from the press and public, while private space efforts push their pilots and inventors to new heights. In short, there are some gleams of hope, surrounded by a lot of the same refrains we've heard for decades.

The US Government Space Program

When people think of the US Government's space program, they think of the National Aeronautics and Space Administration. There are other portions of the US space effort, but most of them are concerned with civilian infrastructure (such as NOAA [The National Oceanographic Atmospheric Administration] and their weather satellites) or military utility that sometimes have civilian applications (an example here is the US Air Force's Global Positioning System, or GPS, which will tell you your latitude and longitude anywhere in the world, as long as you have the proper equipment). Since NASA is the focus of the most interest, we'll spend most of the time talking about them.

Organization

The US space program is divided between the civil side (where the lion's share exists in NASA) and the military side where the majority is funded through the Air Force, although other entities have large, and classified, budget levels. While NASA missions get all the headlines, the military side of space actually launched more satellites throughout its history. It could also be argued that military space has more of a direct impact on our lives.

NASA's organization is made up of rather loosely grouped field centers that all answer at various levels to headquarters in Washington, DC. The loose grouping comes about because very few of them were re-named as NASA centers when the organization was founded. Most of them existed before as military bases, and were then folded into the newly formed NASA by the National Aeronautics and Space Act of 1958. Two NASA centers that were created as part of the initial organization were Goddard Spaceflight Center in Greenbelt, MD, and Johnson Spaceflight Center in Houston, Texas. Kennedy Space Center in Florida, where the space shuttle flies from, was actually split off the Cape Canaveral Air Force Station, and Kennedy only has two launch pads. Any launch from the United States that is not the space shuttle, taking place out of Florida, is happening on an Air Force Base. Other centers include Marshall Spaceflight Center (a former Army Base), Dryden Research Center (a portion of Edwards Air Force Base), Langley Research Center (Air Force), The Jet Propulsion Laboratory (not formally a part of NASA, though they keep the acronym as part of their web address. This lab is part of The California Institute of Technology.), and Ames Research Center (formerly Navy).

Even though these centers all have the same crest on their entryways, much has been made of the fact that there was no single identification badge that would get you into all of them. Someone with a Goddard ID could go to Dryden and they'd be treated as though they walked in off the street (that's quite a feat at Dryden, by the way. That place is in the middle of a desert—pretty desolate.) Before being known as the second NASA chief to lead the organization during a shuttle disaster, current administrator Sean O'Keefe was probably best known for his "One NASA" initiative, trying to get all centers to recognize each other's badges. Progress on the effort has not been reported widely in the press.

While each center has a specialty, Johnson for crewed spaceflight, Marshall for propulsion, etc., there are also overlap areas and these are hotly contested. Whenever budget axes fall, the sibling centers bicker over the remaining funding. Typically, this is where a strong headquarters function would kick in, telling the siblings to shut up and take the money that they've got, but NASA centers have powerful friends in Congress, and in many cases, the money in NASA's budget is routed directly to the centers as part of the budget process. So this leaves HQ somewhat hamstrung, with less authority than they would have if they controlled the money directly. This type of budget is difficult to manage, but it's not very much different than what happens to other government agencies.

NASA headquarters is organized as a series of codes. Each code is given a letter, and to hear NASA people talk is to hear an alphabet soup of where people work with very little additional information. As with most organizations, once you're familiar with the code system, it becomes second nature. Some examples of codes include: code S (Space Science), code R (Aeronautics), code M (Space Flight), and code T (Exploration Systems).

Budget

No matter how much press a NASA budget gets, in modern times the dollar figure has never been over 1% of total US Federal spending, which comes as a surprise to many people. There were a couple years where NASA's budget exceeded one cent on the dollar of the federal budget, but they were in the early ramp-up days to the Apollo program. After those years, funding dropped below 1% and has stayed there.

In 2005, the presidential budget of the United States requested $16.2 billion[1] for NASA. This reflects a 5.6% increase over 2004, and is out of a total US budget of $2.4[2] trillion. This works out to be about .675% of total spending. Now, through the budget process, this number will be debated, have amounts added and subtracted, and be reported on extensively before it finally becomes law sometime down the line when the US Congress actually approves it. It's likely that in the final approval, several projects that were not asked for will be added to the budget and several projects requested will be deleted. These changes are called earmarks, but the added projects are less flatteringly known as congressional pork.

In the past, NASA money has been divided between a number of enterprises, and while the practice is likely to continue, the numbers in each column are likely to change, allowing the agency to take on President Bush's goal of sending humans back to the moon and on to Mars. The typical division of funds in the past had 39% of NASA's budget go to crewed space flight (the space shuttle and the ISS, soon to be joined by the Crew Exploration Vehicle [CEV] and other efforts), 26% for space science, 11% on technology, 10% for Earth science, and 14% for "other"[3].

These numbers seem large, but in President Bush's budget request for 2005[4], there are many other segments of the budget that are larger. For example, defense requests are $401.7B (separate from Homeland Security), Health and Human Services comes in at $66.8B, The Department of Education weighs in at $57.3B, and The Agriculture Department crosses the line at $19.1B. They will be subject to the same debate, and some amount of pluses and minuses, but won't receive the media coverage that the NASA budget does.

Headliner Projects

Headliner projects for NASA are the space shuttle and the International Space Station (ISS), followed closely by their robotic space explorers and the Hubble Space Telescope. Headlines have little to do with returns in some areas, however, because the HST was rated as NASA's number 1 project on science return. Scientific breakthroughs in astronomy rarely make it to the front page of a newspaper, and are often relegated to back pages of print or special interest sections of broadcast news.

The space shuttle is grounded as of now, due to the *Columbia* breakup over Texas in February of 2003. This leaves the space station in a sort of limbo, unable to continue construction (all new pieces made by the US were designed to fly on the space shuttle only) and unable to support more than two people at a time. The second fact is caused by loads of water and other supplies that the space shuttle normally brings up to the station, which won't happen again until after the problems that led to *Columbia*'s loss are corrected.

Luckily for grabbing headlines, a number of uncrewed spacecraft were either ready to go or already in flight when the shuttle accident took place. NASA has a spacecraft en route to Saturn, called *Cassini*, which arrived in July of 2004. The Cassini mission carries a probe on board that will enter the atmosphere of one of Saturn's moons, Titan. The orbiter is equipped with an outstanding camera and other scientific instruments that will provide some amazing images of Saturn the planet, its rings and its moons. When compared to the *Galileo* spacecraft orbiting Jupiter, which was hampered from achieving its full capabilities by a fouled transmission antenna, the number of images from Saturn may be overwhelming.

Another mission that caught some people's interest during the shuttle downtime was Stardust. The goal for this spacecraft was to fly close by a comet, and capture some of its tail for return to Earth. This will be the first time material from a comet was purposely captured and brought back. Of course, to make that statement you have to assume that the moon rocks brought back by Apollo astronauts and Russian robots are not interplanetary samples. The cometary material captured by Stardust will return to Earth in 2006[5], aboard a capsule landing in the Utah desert. A lot of scientists are very excited to see the analysis of the material inside Stardust right now, because most believe that comets are building-block remnants of our solar system.

The Hubble Space Telescope has been the crown jewel of NASA's astronomy efforts. Designed and built as an upgradeable and repairable piece of hardware, Hubble stumbled a bit when it was first deployed due to an improperly ground mirror that drew lots of news stories right after its delivery to orbit. Because of its upgradeable nature, the space shuttle was able to travel to Hubble with a group of astronauts who repaired the telescope and brought back its promise. Since the servicing mission in 1993, the Hubble has stunned the world with beautiful images from the very edge of our universe. One primary goal of the telescope was to provide information for calculating the age of the universe, using the methods proposed by its namesake, Edwin Hubble. Using that methodology, and the data supplied by the orbiting observatory, scientists have estimated our universe's age at 11-14 billion years.

At this writing, there are those who question Hubble's future. Soon after President Bush's new space initiative was announced on January 14, 2004, NASA cancelled the last servicing mission to Hubble, citing safety concerns as part of the space shuttle's return to flight. As discussed in later chapters, the Columbia Accident Investigation Board (CAIB) recommended that the space shuttle be fully self-supporting in relation to events that caused *Columbia*'s breakup—being able to diagnose and repair any damage to its outer skin. According to NASA today, the only mission where such diagnostics and repair would be necessary is the Hubble servicing mission, because all other shuttle flights on the books will go to the ISS. This argument is not air-tight, because the CAIB recommends a self-diagnosis and repair capability on all shuttle missions, not even those to the ISS. The board's reasoning is that some missions to the ISS may not be able to reach their goal due to problems en route, and the orbiter on that mission can't rely on ISS for diagnosis and repair.

The combination of this uncertainty with the outstanding science return that Hubble has given us during its years of operation has caused a groundswell of laypersons, scientists, and politicians to rally around the Hubble to try and stave off its abandonment in orbit. Through websites[6] and letters, the interested parties want to see the last servicing mission reinstated, to keep Hubble operating longer, hopefully until the next generation space telescope can fly. This opposition is only a peek into the complicated world of space policy.

Ongoing Research

Research at NASA is roughly divided by its organizations, and deals with human spaceflight, space sciences, Earth sciences, and aviation.

Human spaceflight takes up the majority of the budget at NASA. This budget is split between flying the space shuttle and building (on Earth) and assembling (in orbit) the ISS. The primary reason for building the station, as stated for quite some time, is to learn about impacts to the human body from long-duration exposure to the space environment. Other research cited on board the ISS is materials science, such as processing new medicines and computer chips. The amount of research feasible on the station has always been difficult to quantify, because of the decreasing number of crew aboard (original plans called for 7, then 3 became the norm, and now with the shuttles grounded only two can call the station home for long stretches). There are other considerations as well. It is impossible to build a space station that excels at all capabilities. The reason is that different processes (learning about humans' adaptivity and metallurgy, for example) require different things to be useful. Human adaptation work requires lots of physical exercise, which means that people are moving around all the time. In space, any person moving on board the station causes the station to move in the opposite direction. Even though such motion is unnoticeable to the crew on board, any ultra-pure medicines or computer chips in production will notice the movement, likely messing up one portion of the research or manufacturing. Of course, there are ways around this: the crew could stay extremely still during some material processes, but that cuts into time for the crew to do other things.

Space sciences deal with learning about the universe around us. Telescopes such as The Hubble Space Telescope fall into this category, as do the robotic explorers sent to other planets to learn about them and report back to us. Robotic exploration is largely focused at The Jet Propulsion Laboratory in Pasadena, California, and telescope efforts are big at The Goddard Spaceflight Center in Maryland. Telescopes are the kind of missions that work in the background, building scientific knowledge over the years as different astronomers from around the world ask for time on the instrument, then take the data and write up their results. This leads to sporadic announcements of particularly amazing discoveries, or of pictures that really dazzle. Most robotic explorers are more short term, either flying by a planet or other planetary body, taking some quick pictures and beaming them back to Earth, or landing on the surface of such a body, sending back images or data for a time before the mission ends. Some missions go on for a long period of time, such as the *Galileo* spacecraft, which orbited Jupiter for eight years or the Near Earth Asteroid Rendezvous (NEAR) mission that circled an asteroid for over a year. For these craft, there's an initial flood of news, followed by a time of obscurity like the earlier-mentioned telescopes. For "pure" science, and those who claim to follow it, NASA's space science mission is truly where money is well spent.

Earth science is another area that NASA focuses on. Using Earth-orbiting spacecraft, the agency looks at our home planet to learn more about it. In this role, NASA conflicts a bit with NOAA, the National Oceanographic and Atmospheric Administration. NOAA is tasked to provide weather data that people see on the news, while NASA's job is to research instruments that will provide new types of data. In a rather confusing arrangement, NASA is also the primary purchaser of civil weather satellites. NASA purchases them, tests them, has them launched, checks them out in space, and then hands the satellites over to NOAA for operation. NASA has some of its own exclusive Earth-science satellites, however, primarily found in the Earth Observation System or EOS. These satellites have instruments on board that are more advanced than everyday weather satellites, and are designed to prove what more capability would do for weather prediction, as well as provide randomly announced datapoints for global warming debates and ozone depletion news stories.

Long considered the forgotten arm of NASA, even though the first "A" in the acronym mentions it by name, is aeronautics. This role for NASA translated from their earlier incarnation as the National Advisory Council on Aeronautics (NACA), which was founded in 1915 and existed until the 1958 National Aeronautics and Space Act folded it into the new organization. NASA's aeronautic focus has been on advanced technology, largely through cooperative projects with the Air Force. Recent successes in solar-powered flight, as well as previous research in uniquely shaped aircraft (such as the X-29 with forward-swept wings) are hallmarks of NASA's legacy in aeronautics. There are no commonly stated long-term goals for NASA's aeronautic division, though 2005 budget charts show a small growth in its budget as NASA's budget grows.

Future Plans

NASA's future plans, such as they existed, were turned on their heads on January 14th, 2004 when President George Bush went to NASA headquarters to outline a new long-term goal for the agency. The announcement was one of the recommendations of the Columbia Accident Investigation Board, in setting a long-term plan for the agency. In the board's opinion, NASA functioned best when it was given a focused goal to meet. The President's goal will be discussed in more detail in a devoted chapter, but for now the main idea is that NASA's job is to move humans into the solar system. As part of this new agenda, the President called for all NASA programs to be evaluated for their support of the stated goals. Any programs found not to support those goals would be cancelled and provide the funding for early research into human spaceflight beyond low Earth orbit. After international treaties are fulfilled with the ISS, that activity as well as shuttle flights will be ramped down, and the funding currently in those programs will be channeled into the new exploration effort.

Right now, NASA is in the midst of a large reorganization to take on the new challenges outlined by The Bush Administration. Time will tell how the new subset of America's citizen space organization takes on the tasks it's been called to do.

THE RUSSIAN GOVERNMENT SPACE PROGRAM

The Russian space program is not nearly the dynamo of activity that it was before the end of the cold war. During that time, the Soviet Union dominated launch rates, sending fully 68% of the payloads lifted from our planet into space[7]. Part of this amazing launch rate had to do with their spacecraft, in that they did not last as long as other nations'. Another factor, though, was the ruling Communist party's view that space was a place to showcase national prowess.

Since the breakup of the Soviet Union, the formerly grandiose space effort suffered from a severe lack of funds. In fact, the Mir space station was occupied throughout the Soviet breakup by a single set of cosmonauts. They returned to a different political reality. Fears grew within the United States that Russian rocket scientists, out of work or just bored, might travel and share their knowledge with nations that weren't friendly to the US. In order to keep the space industry busy in Russia, the idea was hatched to ask Russia to participate in what now is called the International Space Station, which was having its own severe budgetary problems. Russia was given the task of constructing two key components of the space station, one of which was paid for by the US and the other was paid for by Russia. Other tasks that Russia would fulfill as a station partner were to provide an emergency space vehicle for the crew in the early stages of station construction, and provide a consistent flow of supplies to the station. These ferry tasks were to be carried out by the *Soyuz* and *Progress* vehicles, respectively. These roles are set to expire in 2006[8], at which point it was supposed that some other form of transportation would be ready to take on the burden, though none is currently planned that will meet the 2006 timeline.

Support for the ISS has become the majority of the Russian space effort. It appears as though there isn't interest in doing much else, largely due to funding constraints. The last interplanetary mission launched by the Russians was the ill-fated Mars 96 probe that didn't make it out of Earth orbit, falling back into the ocean soon after launch. Military space efforts continue, though at a much lower level than before the collapse of the Soviet Union. The decay of the Russian military space infrastructure has led to at least one scary moment bringing back the days of the Cold War, where a rocket launched from Norway was mistaken as an American attack[9]. For a short period of time, President Boris Yeltsin weighed his options for a response while the whole thing was sorted out. Needless to say, that tale did not end in the destruction of humankind, or the publication of this book would have been delayed.

So Russian space efforts today are limited largely to servicing and transferring crews to the ISS. These roles became more important after the *Columbia* disaster.

While providing this service, Russia has gone rather maverick in their approach, and has actually turned its station taxi service into a revenue source. The Soyuz taxi flights are programmed to carry two people up to the station and then typically (when the shuttle is flying) bring the same two people down. Well, the *Soyuz* has three seats, and Russia, with some booking help from corporate start-ups in the US, sold a seat on two of those missions. One seat went to an American businessman Dennis Tito, and the other went to a South African named Mark Shuttleworth. The fee they each paid (rumored to be $20M, though the details are closely guarded) included their training and a flight up to the station on one of the Soyuz swaps. Each passenger got a chance to be on board the ISS for about a week, and while Dennis Tito describes just looking out the window and listening to opera, Mark Shuttleworth actually took some scientific experiments up with him and carried them out. Very little has made it into the popular press about the results, but this is not a good gauge of science experiments in space.

This effort appeared to catch NASA off-guard, and the agency had no action plan for Dennis Tito's flight. Press conferences up until days before launch showed a shocked organization, with repeated statements like "We expect Russia to act like a partner in this project" giving hint as to how much pressure was applied behind the scenes to prevent Dennis Tito from getting his ride in space. There was even a standoff of sorts in Houston, where NASA refused to train Tito on the American portions of the station. In defiance, Tito's Russian crewmates also refused to train. In the end, NASA had the first tourist (and the Russian government) sign some papers saying that they wouldn't break anything, and Dennis Tito flew. Very little was publicized of the flight so it passed by in relative obscurity compared to what it could have been, and in my opinion, that mission was the biggest space "first" in years.

Recent news articles[10] state that Russian and its American corporate partners will start flying paying customers along with their Soyuz flights again in late 2004 and 2005. An unstated assumption would seem to be that shuttle flights will resume first, but shuttle flights were not necessary for three people to fly to ISS. In October 2003, a three-crewed mission was flown to swap out crews and the old *Soyuz* vehicle. The third crewmember remained on board for a week, and then went home with the crew being relieved from the station.

Russia has gone commercial in some of their launch services as well. Treaties with the United States require Russia to dispose of several types of missiles, and the Russians realized that it would be more economical to sell the missiles as rides into space than to smash the empty hulks with a bulldozer. One agency that's taking advantage of the opportunity is The Planetary Society, which is asking Russian space companies to build a solar-sail spacecraft and launch it on one of the surplus rockets.

Of course, space advocacy groups who favor a more commercial approach to spaceflight love to point out that Russia, the former communist nation, has a more commercial approach to their space program than the capitalism-based United States. While this is true, the capitalistic efforts within Russia haven't yielded enough cash to keep their program going yet, and I haven't read about any serious business plan that would take them there soon. Still, there's something to be said for Russia's records: the first nation to launch a person into space and the first nation to launch a paying customer into space.

EUROPE

The European Space Agency (ESA) is made up of individual countries' space agencies from across the continent. The primary member, as shown through budget levels, is the French space agency Centre National d'Etudes Spatiales, CNES, followed closely by Germany and more distantly by Italy, but other countries contribute, normally through their own projects. The total budget for ESA was 3.1 billion euros in 2000, split among a number of projects.

The primary effort that ESA is known for is its Ariane launch vehicles. Launched out of French Guiana, South America, this series of vehicles take advantage of an excellent launch position on the Earth to lift communications satellites into orbit. The Ariane 4 rocket was the surprise (at least to the United States) commercial hit of the space age. Whereas the US launched the majority of communications satellites for the world in 1980, Europe launched over 50% of the commercial satellites in the world at the beginning of the 21st century[11]. These stunning results prompted Europe to research and build the Ariane 5 rocket, a much more powerful booster with very little in common with the Ariane 4. The new booster has suffered some spotty performance numbers, currently holding a success rate hovering near 90%[12] but it remains the booster of choice for many of Europe's space projects.

Europe is a participant in the ISS, and it is in the process of building a module that will attach to the orbiting outpost. Called Columbus, the module was originally hoped to be free flying. This free-flight capability would be much better for manufacturing experimentation, but the costs of making the facility fly on its own drove ESA to attach Columbus to the core station. Another major contribution that Europe is working on is the Automated Transfer Vehicle, or ATV. The spacecraft will provide cargo capacity between the Russian *Progress* spacecraft and the American space shuttle. The uncrewed ATV will lift off on top of an Ariane 5 vehicle, then fly to the space station with its cache of supplies. After docking with the station, the ATV will be able to use its own engines to push the station to a higher orbit. This is required because the ISS constantly loses altitude through friction with the upper atmosphere. After the crew unloads the ATV's supplies and it does all the orbital maintenance it can, the vehicle separates from the station and burns itself up in the atmosphere. Some literature has mentioned a Crew Transfer Vehicle, or CTV, that would carry four crewmembers to the ISS, and also provide additional emergency escape capability, but news reports on the vehicle have been nil.

Uncrewed science missions have also been a priority within ESA. Most of the early efforts were joint missions with NASA, but ESA is moving out on its own in some exploration. Some examples of missions in flight include *Ulysses*, a spacecraft designed to explore the sun's polar regions, and the Solar and Heliospheric Observatory (SOHO) to monitor the sun from an excellent vantage point between the Earth and its mother star. ESA also built the Huygens probe, which, at this moment, is flying to Saturn along with the

NASA Cassini mission. After arriving at Saturn, *Cassini* will release Huygens to explore Saturn's moon Titan. It will be the most in-depth analysis of another planet's moon ever done. The SMART-1 mission to our own moon is taking a very long, but very fuel-efficient route to our nearest neighbor by using advanced propulsion to push it on its way, and at this writing the Rosetta mission is preparing to fly to explore a comet.

The Mars Express mission made headlines in late 2003, focusing largely on the failure of its landing craft, named *Beagle 2*. *Beagle 2* flew piggy-back on the *Mars Express* spacecraft, and was dropped off just before final maneuvers were made to steer the mothercraft towards orbit around Mars. If all had gone well, *Beagle 2* would have bounced to a stop on the surface of Mars and used its amazingly small yet capable science lab to look for life on and under the surface. All did not go well, and because of its mission design, we have no information about what went wrong. The orbiter, meanwhile, entered Mars orbit with no problems, and used its camera and radar systems to probe the planet's secrets. Recent press announcements[13] state that the craft confirmed earlier US findings that there is water ice on the surface of Mars, and because of the methods used to investigate, the European information is more solid than the US's. Mars Express is Europe's first solo exploration of another planet, and ESA is rightfully quite proud of the achievement.

ESA dismayed the United States somewhat when they stated that they would pursue their own global positioning system, which they call *Galileo*. Developing an indigenous navigation capability would prevent Europe from relying on the US's Global Positioning System constellation. I've not read any reason why relying on the US is a problem, especially considering the close alliance between the nations through the North Atlantic Treaty Organization (NATO). The biggest complaint that the US had about the system was that the *Galileo* transmission was going to interfere with US navigation satellite signals, though officials were tight-lipped about the problems this would cause. Later negotiations led to an amicable solution, and current Galileo plans will not "step on" the US frequencies.

Japan

The Japanese space agency was once separated into three organizations: The National Space Development Agency of Japan (NASDA), The Institute of Space and Astronautical Science (ISAS), and the National Aerospace Laboratory of Japan (NAL). An acquaintance of mine was fond of pointing out that competition between these three organizations kept the Japanese space effort much leaner and meaner than other national space programs. Apparently, the Japanese government disagreed, and the agencies are in the process of merging into a new agency called the Japanese Aerospace Exploration Agency (JAXA)[14]. Total expenditures on space efforts in Japan were listed as 207 billion yen in 1998, although NASDA got the lion's share of the funding, coming in at 185 billion that year, or nearly 90%. Funding has been cut in recent years[15].

The Japanese space program has focused recently on building its indigenous launch capability, through its H-II and H-IIA boosters. The H-IIA booster has not had a good track record, and is currently grounded while the space agency troubleshoots problems. This delay in launches has turned into more than an inconvenience for the Japanese. One example of the problems created is that their primary weather satellite is failing, which puts them in danger of having no weather data at all. To maintain some weather information within the hemisphere, the Japanese signed an agreement with the United States and paid for construction of additional equipment (antennas, computers, and the like) that allows the US weather agency, NOAA, to fly one of its older satellites over Japan. This alleviates the immediate need for a weather satellite launch.

The Japanese are also players in the ISS, and they are building the Japanese Experimental Module (JEM) that will be attached to the station later in construction. The important research tool scheduled for inclusion in the JEM early on, but that's now in flux, is the centrifuge module. A centrifuge would allow experiments that expose Earth life to varying gravity levels. That's important because we know that microgravity (what space explorers face when they're onboard the station) is debilitating over time, but we don't know the effect that partial gravity such as the moon's (1/6 of Earth's gravity) or Mars' (a little more than 1/3 of Earth's gravity) will have on people or animals.

CHINA

The Chinese space agency, a fairly secretive effort closely tied to the People's Liberation Army, has worked to build a launch capability for China using lessons learned from ballistic missile development. This is not rare, as most countries have done the same thing. The Chinese launch services program suffered some very public failures, however, in one case decimating a town with an errant rocket.

China's most recent news event in space, however, was its becoming the third nation, behind the United States and Russia, to place a person in orbit. The flight took place in October of 2003 and was successful, with the pilot enjoying the trappings of instant celebrity, much the same way as the first astronauts in The Soviet Union and the United States did.

China's spacecraft are based heavily on Russia's *Soyuz* capsule, though the Chinese made some advancements that prove they are not simply copying what someone else has done. *Soyuz* vehicles have a round segment at the front of the capsule, called the orbital module. On the Russian version, this orbital module separates from the rest of the vehicle when it's time for the crew to return home and burns up in the atmosphere on the way back to Earth. China altered the orbital module, adding power-producing solar arrays. This allows the crew to leave the orbital module behind, where it can fly on its own for some time. Presumably, future missions will travel to the same orbital module and dock with it, and a string of these modules, which others have simply thrown away, may provide a backbone for a small space station.

China's news agencies report that plans include more flights into low Earth orbit, followed by eventual exploration of the moon. Some space pundits point to these statements and say that China is a new threat in space, and should be treated much the same as the United States' cold war adversary in the '60s. I disagree. China took several years of very methodical testing before they proceeded with their first flight. It's unlikely that a space agency, flush from its first success, would throw out its test plans for an immediate leap for the moon after just a few flights in orbit. I am very curious to watch Chinese plans unfold, however.

INDIA

India has the capability to launch its own satellites. As with most nations, the rockets are based on their ICBMs. India also has an industry to build satellites, and has expressed an interest in sending probes to the moon.

Treaty Organizations

Alliances of nations allow those without the resources for their own space programs to participate in, and receive the rewards from, such activities. To date, most of these organizations are focused on providing communications services to their member countries, but there's no reason that they couldn't take on a more active role in space much like ESA does.

Space Prizes

The X-Prize[16], unveiled in the mid 1990s, offers $10M for the first group of people who fly three people to 100km in altitude and then repeat the flight using the same hardware (not necessarily the same people) two weeks later. Today there are 27 teams signed up for the X-Prize. While some of the teams are apparently made up of one person with unknown skills or resources, others come from relatively well-known aerospace designers.

Space prizes seem like a new idea, but they are merely an upgraded version of the aviation prizes offered for aircraft firsts in the early-mid 1900s. Charles Lindbergh was competing for just such a prize in 1927 when he flew across the Atlantic Ocean solo in *The Spirit of St. Louis*. By the way, the X-Prize has tapped into that history, by basing most of their fundraising effort in St. Louis, and asking that participants in the competition name their craft *The New Spirit of St. Louis*.

The X-Prize was won in October of 2004, when Burt Rutan's *SpaceShipOne* flew for the second time in five days to the necessary altitude with the necessary mass payload aboard. Because of licensing issues, the craft could not carry anyone other than the pilot, so Burt's company, Scaled Composites, filled the back two seats with personal memorabilia. This event sparked a flurry of activity, from the announcement of Virgin Galactic, a space tourism company sponsored by Richard Branson, through new X-Prizes designed to bring about technological and social change.

This comes as a relief for me, because I was peripherally involved in another space prize effort, known as the CATS (Cheap Access To Space) prize in the late 1990s. The Space Frontier Foundation announced this prize in 1997 at their annual conference. Here, the prize was much smaller ($250K), but the goal was set lower as well. To win the CATS prize, a participant had to send a 2kg payload chosen by the Foundation to an altitude of 200km, using hardware that was developed almost exclusively from scratch. Despite a lot of bravado that I saw at the announcing conference, with people claiming that they'd win the prize in six months through sponsorship by documentary TV shows, the prize went unclaimed. To my knowledge, there was only one serious attempt to actually claim the prize, a launch by a group known as HARC, but that flight failed[17].

There are many within the space advocacy community that believe that larger space prizes, offered by groups such as the US Government, would open a new chapter in affordable space activity without repeating mistakes of the past. In recent space development projects, government agencies embark on grand projects, promising either to revolutionize space travel, or demonstrate that such a revolution is possible. As problems mount for the program or priorities shift within the government, those projects are cancelled, but the money ($5B according to a *USA Today* article) to develop the system has still been spent. The years 1986-2003 may hold those failures as America's space agency's primary legacy[18]. Granted, there are some benefits from research for research's sake, but the returns on a partially built spacecraft are debatable at best.

The space prize argument goes that prizes will prevent this sort of wasteful spending, because a stated goal will be put out with a dollar figure attached to it, and the first person to achieve that goal will win the money...there will be no infighting about capabilities or congressional districts represented.

The government-offered prize idea sounds good on first blush, especially since it was successful in the past, but there may be some problems in its execution. First, it's the government's role to look out for the public good, but the primary method of doing that is by controlling things. A prize format does not allow any control on the actual procurement. Why does this matter? Well, when it comes down to brass tacks, a congressperson's reelection typically doesn't come down to how well he or she prepared the nation for the future, it relies more on how many jobs were created or lost in their district during their term. If Congress voted to give a prize to anyone who could develop a new launch vehicle, and the winners of that prize turned out to be from, let's say, North Dakota, the congresspeople from traditionally known space areas will have hell to pay. Even if the new vehicle did not threaten current launch vehicles directly (let's say it was a large booster, designed to lift Saturn V moon rocket-class payloads, so it would be impractical for lifting small satellites like the rest of the current rocket fleet) there is an entire group of lobbyists on standby in Washington, DC waiting to point out how the new vehicle will threaten jobs in California, Texas, Florida, and elsewhere. Any lobbyist that's worth their salt would head the North Dakotan development effort off at the start.

Another thing to remember is that all the bluster aside, Congress' job is to control things. Under the United States Constitution, the primary method that they use to exercise that control is the power of the purse, in setting budgetary levels for government spending. Here, Congress relies on the fact that the budget is so complicated that few in the general public are interested. Where else but in US Government dealings could two members of the House of Representatives have a ten-minute debate (discussion/argument) about whether or not an organization's budget has gone down or stayed the same? With the public's lack of interest taken for granted, then maneuverings within particular programs go on all the time. In a fairly well-publicized case, the Pluto-Kuiper Express mission (the first planned mission to visit Pluto and a body beyond), NASA (and, by extension, the President) essentially cancelled the mission by not requesting any funding for it. Over the years, though, Congress would add funds to NASA's budget, specifically set aside for that mission, and at this writing it looks as though the craft will fly. Although fuel processing problems (brought about by security issues at Los Alamos National Laboratory) may have the craft launch with less nuclear fuel aboard than necessary. This may cut the mission short. While many consider this example a positive one, there are plenty of cases where the same type of power was used to cancel some very promising technology development.

How does this all play into space prizes? Once Congress goes on record stating that the winner of a prize will receive some amount of money, that will be the end of their involvement. I believe that fact will drive them crazy. By definition, a prize competition will be developed completely out of government control for capabilities, flight rates, production location, and the like, and this is the kind of stuff that Congress loves to meddle in. What if the prize year approaches, and there's some new hot-button topic that needs to be funded? Will the funds be taken from the prize, just as it's about to be won? A company or private interest that's put a lot of time, effort, and resources into a solution may not take too kindly to having their prize reduced at the last moment. If the prize went unclaimed at that point, whom could the lawmakers blame? At that point, it's likely that they'd take the "well, we never spent the money anyway" approach and spin things just so that it still looks like everyone did everything right. This would take a minor amount of spin compared to most items in political circles.

As with many positions I take within this book, I want to emphasize that I believe prizes (or, perhaps written more palatably, delivery-on-orbit contracts) have a tremendous capability to revolutionize our approach to space, but not without risks and outright down sides.

There's another difficulty this type of acquisition faces, in that it's never been done before on a large scale. It is true that the Defense Advanced Research Projects Agency (DARPA) announced the Grand Challenge in 2003[19], but this prize is only for $1M, a very small sum compared to those required to inspire space-interested prize seekers. The actual contest, held in March of 2004, had its best vehicle travel about 7 miles, far short of the 142 mile goal. Despite this initial performance, America's space agency appears to be waking up to the prize concept, announcing Centennial Challenges, where NASA offers prize money for innovation in space. I attended its first workshop, and a lot of people in attendance were excited at the prospect, but funding problems in Congress may lead to trouble.

"Pure" Entrepreneurships

SpaceX is short for Space Exploration Technologies. This company was founded by internet millionaire Elon Musk who moved his efforts into space travel after founding, making successful, and then selling two internet companies at a tremendous profit.

SpaceX's primary goal at this writing is to develop and launch a commercially viable small satellite rocket, named Falcon, capable of delivering about 450 kg (1000 lbs) into low Earth orbit. The booster is a relatively simple two-stage design with both stages being powered by kerosene and liquid oxygen. Its first stage is designed to be recoverable, although the practicality of this is not known at this time. I saw Elon speak at a conference, and his straightforwardness about the first stage's reusability was refreshing, since I come from an aerospace background where lots of careers are made pounding on podiums in meetings saying that some untried idea will work. In essence, Elon is saying he hopes that the first stage will turn out to be reusable, has simulations saying that it will be, and can't wait to see whether or not it actually is. First stage reusability will bring the re-use mass of Falcon to nearly 80%, which compares well to the space shuttle's 90%, and comes in a much smaller, less expensive package. In their September newsletter, SpaceX announced the selection of a sea-based salvage team that will bring their used first stages back.

Future versions of the Falcon will carry more payload. The second Falcon, announced at a glitzy unveiling of the original booster on December 4 of 2003, is called the Falcon V [20]. The name comes from the fact that five Falcon first stage engines will be attached to a new first stage to provide thrust for the vehicle, but it's also likely meant to be a tribute to the Saturn V, the rocket that took people to the moon in the late '60s and early '70s. The announcement stated that the Falcon V would have engine-out capability, that is the booster would be able to continue on its mission even if one of the its engines failed. This capability hasn't really existed since the Saturn V days, and according to SpaceX, allowed two lunar missions to continue despite one engine's failure in flight. The critical point, though is that the Falcon V will be able to lift satellites weighing 4200 kg to low Earth orbit, and be able to send 840 kg on an interplanetary trajectory. An additional, higher-performance Falcon V has been discussed in magazine articles costing $20M and lifting over 9,000kg to low Earth orbit.

The flight into orbit on board Falcon costs $6M, and that price assumes that first stages can't be reused. When Falcon V starts flying, it will cost $12M. These numbers are unheard of for a space launch. The closest competitor for the first Falcon, the winged Pegasus rocket built by Orbital Sciences, costs between 3 and 5 times as much as a flight aboard Falcon (Orbital says that you can get a ride aboard the booster for $18M, but NASA lists a cost of $30M for a ticket), and its payload capability is lower. The Falcon V's payload approaches that of a Delta II booster, but the cost is 1/5 the amount. In announcing the Falcon V, Elon bragged about a cost per pound to orbit for the first time, and placed the value at $1300/lb. His business plan is to show that low-cost flights like these are possible and lucrative, after which he'll move into higher-power rockets at presumably much lower costs than current launchers.

When asked, Elon says that SpaceX's infrastructure should be able to support up to 4 launches per year as they're made up right now, and that one of the most expensive (and difficult) portions of the work is the flight termination system that's required on board the rocket. Flight termination systems blow a rocket up if it goes off course. Of course, the irony of the situation is that this system is required on any rocket, so it would be the easiest to standardize and make much cheaper, but this hasn't happened in the current launch vehicle environment.

In designing the Falcon, Elon consulted with a report produced by The Aerospace Corporation that said that 91% of launch failures were caused by stage separation problems, engine problems, or guidance problems[21]. With that as a start, he poured the majority of his research into making the Falcon's systems in those areas extremely robust. This effort led to the simple design known as Falcon. It's a bragging point for Falcon that the vehicle's first flight will carry a paying customer's satellite, as recently announced. DARPA's TacSat 1 will ride Falcon into orbit when it launches in early 2005. There are other entrepreneurial players in this mix, however.

A group called Team Encounter has an interesting question: Would you like to know that your DNA has been captured and sent into space? How about a poem or drawing that you did, would you like a digital representation of it to leave the solar system, theoretically to be found one day by an alien species who'll marvel at your genius? If your answer to either of those questions is yes, then they have a deal for you.

According to their online presence, Team Encounter is collecting DNA and electronic works to send into space. To get your material on board, all you need to do is go to their website[22] and purchase a kit. There are a couple of different options, depending on how involved you want to be, but you either send them a hair follicle or some sample of your work. Then, sometime in the future (the company used to be called Encounter 2001, but their launch date slipped beyond their name, so they re-thought the approach) they will bundle up everything that's been sent to them, place it on board a rocket, and then fly it out of the solar system using a solar sail.

I've heard presentations on the Team Encounter business model, and while I'm not sure how viable it is for repetitive flights, for this first mission they may have hit on something. By having interested parties contribute $25-80 for individuals or families to "fly" aboard their spacecraft, they have a fairly early payoff for their development. This is a very novel concept, and will catch the imagination of most of the space community with some other interest as well. Right now, the website states that over 100,000 people have signed up. Now, if the first mission is a success and makes big news, their second mission may be completely overrun by interest from the public at large. From there, I'm not sure how long the novelty will last.

Their launch plan has remained pretty constant, in that they will fly as a secondary payload on board an ESA Ariane flight. The propulsion system that they plan to use once in orbit has changed, with their first plan giving their payload a traditional boost from a chemically propelled upper stage. It was in the year 2000, when their development timeline showed that they wouldn't be able to launch to meet their original name, that they decided to delay the flight and change propulsion technologies, moving to a solar sail.

Solar sails have been a science fiction concept for years in works featured by authors such as Arthur C. Clarke[23], but recently, movies and television have shown examples of solar sails including an episode of *Star Trek: Deep Space 9*, and the George Lucas film *Star Wars: Attack of the Clones*. The concept of solar sails is valid, in that light from our sun exerts a force on anything exposed to it. The problem comes in that the pressure (force per unit area) of that force is extremely low, so anything that wants to use light for propulsion has to be extremely light and extremely large, which brings a sail to mind.

Here, Team Encounter really outdid themselves by pushing development in solar sails ahead by years compared to any government research. Team Encounter pioneered the development of extremely light sail material and structures to hold the sail, catching the interest of two government agencies…NASA and NOAA. This interest drew some money into the project, but also pushed timelines back further, and caused me to soften the title of this section, Pure Entrepreneurships with the quotes around pure.

Another rich internet entrepreneur, Jeff Bezos, the founder of Amazon.com, decided that he wanted to have his own space program too. As I researched this book, Blue Origin was very secretive about their project, although some word has leaked out that they're looking at a vertical take-off/vertical landing vehicle of some type. In this project, there's no lack of funding, and as Jeff Foust says, money is the best rocket fuel[24]. So, with the right talent, something just may come of it.

In the desert of Nevada, just outside of Las Vegas, Bigelow Aerospace is set up to take advantage of commercial space activities. Founded in 1999 by businessman Robert T. Bigelow, the company started out stating that they were going to build orbiting hotels and spacecraft to travel to them. Bigelow Aerospace's web page has recently posted many details about the company's activities, discussing the Nautilus system. This concept capitalizes on early NASA research into inflatable living spaces for use in Earth Orbit, and Mr. Bigelow believes that they will be critical to opening commercial space[25]. Bigelow is contracted with SpaceX to fly a Nautilus demonstrator on board the first Falcon V flight. Then, as the X-Prize flights took place, Mr. Bigelow announced America's Space Prize, with an offer of $50M to any team that will build a craft to fly to his future, inflatable stations. The bigger plum in the offer, however, is the promise that the award winner will be first in line for a contract including regular delivery to Bigelow projects.

Tom Hill

Non-Profit-Sponsored Activities

Mars Research Stations

At the Mars Society's founding convention in 1998, the general membership voted to make the organization's first project an analog research station in the polar desert on Devon Island. The idea of an analog station is that a group of people get together, with suitable equipment, and then they enter the station. As soon as the door closes, for all intents and purposes, they are "on Mars." Any time the people want to go outside, they have to be in a simulated space suit. Any research they want to do will have to be constrained by the fact that it has to be done in a period of time that the suit can support, and they'll also have to work in heavy gloves. This changes things quite a bit from demonstrating an experiment in a pristine lab setting.

Of course, before any research can be done at an analog station, the station has to be built. For something to be built, it has to be designed and paid for, and these were the first issues The Mars Society faced. Design work started immediately after the first convention, as an architect named Kurt Micheels volunteered his services to design the habitat. This is where I got involved in the process, trying to pull together members of the habitat team in the mid phases of the internet boom. Design continued through the first full year of Mars Society existence, with two meetings being held in the San Francisco area, and the design work so far being presented at the Mars Society convention in 1999. From there, the station moved towards construction, with bid requests sent out in 2000 and the goal of constructing the habitat at Devon Island during the 2000 summer season there.

The design of the habitat was constrained by the fact that it had to be moved to the Arctic at some time. It was fun, early in the design phase, to imagine the whole habitat carried intact to Devon Island under a helicopter, and then gently set down. Imagine what it would look like with a properly pointed camera (not showing the helicopter) and a little red filtering! As built, the habitat consisted of 8 sections that formed a ring, and then 8 pie-shaped pieces that came together to form the roof. These pieces would be shipped to the Arctic on board US Marine cargo aircraft and airdropped into position. Assembly problems, common with a project that's never been done before, pushed the deadline and eventually TMS had to enlist volunteer help from the Rocky Mountain Chapter of the Society itself to assist the contractor in tests.

After a test build in Colorado, the station was disassembled and split into drop pallets: the first few carried the habitat outer shell, while the last one carried the floors and assembly equipment. The first drops went fine, but the last suffered a parachute failure and was a disaster. Nothing from the fourth drop was salvageable. Seeing this, the contractor team hired to assemble the habitat left, figuring that the habitat would not be constructed that year. Even the architect who'd worked the project from the beginning left, saying that the equipment on board the fourth pallet would have to be re-built and dropped the following year for a successful assembly.

Mobilizing volunteers and Nunavut natives, and using lumber purchased and shipped in from the nearest town, ropes and manually powered pulleys, a different team than was originally to assemble the habitat put it all together. The habitat was assembled, and there was even time to get a small simulation done. The first analog habitat was complete.

The habitat on Devon Island is populated for about one month a year, when the weather allows such operations. Because of this limited window, The Mars Society looked for another location to place a second habitat, one that could operate for most of the year. They chose the desert southwest of the United States, and contracted to have another habitat built there. An interesting development took place in the funding of the station, because unions of sheet metal workers and plumbers and pipefitters were major sponsors of the station[26] saying that, basically, it is their workers who actually build aerospace equipment once it's envisioned. After initial assembly, but before shipment to Utah, the habitat spent a couple months at Kennedy Space Center, allowing tourists to walk through it. The Mars Desert Research Station is typically crewed in two-week rotations by groups of people interested in the activity. Scientific or engineering background is a help, but is not required, and anyone interested in participating should check The Mars Society website[27] for the latest application information. Plans are in place to put another station in Iceland and the outback, allowing Europeans and Australians relatively easy access to similar opportunities.

There will always be some debate about the practicality of these analog stations, with detractors saying they accomplish nothing and proponents believing that they're doing the best research possible for humans-to-Mars exploration. I believe there's a definite advantage to giving an experiment-proposing scientist insight into how their hardware will function in the real world. Having a pool of people who've been "on Mars" for a couple weeks and can talk about what it's like is also a plus. I hope to participate one day in their research. One of the original arguments for this project, that allowing the public to see what Mars exploration looks like will make it seem more realistic in their minds, is also good, but coverage of the activities is spotty at best, even in the science-devoted press. Perhaps some "flashier" work being done at the stations would improve that image, but flashier work typically has less substance, so a trade will have to take place.

Cosmos 1

Cosmos 1 is currently scheduled to be the first solar-sail spacecraft to fly when it lifts off in 2004[28]. Solar sail technology, as described earlier, deploys very large, but very light solar sails that react to the sun's light pressure. That light pressure, though it produces a very small thrust, is constant, and allows a solar sail craft to do some amazing things. Designed by The Planetary Society and built in Russia, it will fly into LEO on a retired Russian missile fired from a submarine. Once in orbit, the solar sails will open, increasing the craft's surface area by hundreds of times. Four of the eight panels can change their direction, and will be used to increase and decrease the amount of solar force on Cosmos 1 in different parts of its orbit. By doing this, the small satellite will increase its altitude. According to press releases, the satellite will be very bright in the sky because of its reflective properties; people on the ground should be able to spot it easily.

Chapter Two

How We Got Here

Lineage: 1000-1957

Rocketry, touted as a modern-day invention, actually has ancient fundamental roots. The Chinese attached paper tubes to their arrows to create the original shock and awe campaign in the year 1232. The noise and smoke that these super-arrows created so surprised the invading Mongul army that the invasion was repulsed with very little actual combat.

Solid-propellant rockets were gradually improved as time wore on, finding use as naval weapons and in land battles as a complement to cannon fire. Later, as artillerists improved their art, and the types of artillery increased, the rocket fell from grace as a primary weapon. Their most memorable use was likely in the shelling of Fort McHenry by the British during the war of 1812. While negotiating a prisoner exchange, and watching the battle from on board a British ship, Francis Scott Key wrote the poem that when combined with an old English drinking song, became "The Star Spangled Banner," the national anthem of the United States.

Solid propellant rockets played a central role in a moon mission design done by the British Interplanetary Society. This group of early space travel enthusiasts (whose membership included a name that most readers will recognize in Sir Arthur C. Clarke) strapped hundreds of fanciful solid rockets together in stages to allow someone in a space suit (described as a suit with an extremely tough exterior, perhaps made of leather) to travel to the moon.

Solid propellant rockets have a lot of useful properties, such as high power and low maintenance, but they also have some drawbacks. A solid rocket cannot be turned off once it starts firing—the only thing that stops a solid rocket on a good day is the end of its fuel supply. A solid rocket also follows a very strict throttle setting (imagine having to choose between "on" or "off" for your car's gas pedal. I have some friends who drive that way, but I prefer to set something in the middle.) It also carries all of its propellants inside, making it relatively heavy and dangerous to handle. In order to get serious, flexible propulsion, another form of rocket is required.

Konstantin Tsiolkovsky was a Russian schoolteacher who dreamt of traveling to the stars. His drawings and theories contributed much to rocketry. He first envisioned the idea of using rocket stages (having large, lower parts of a rocket fall off while other, lighter parts continued on) to improve performance, as well as the concept of using liquids to power a rocket. In a liquid-propelled rocket, the fuel and oxidizer (which can be as simple as kerosene and liquefied oxygen) are held in separate tanks until its time for them to do their job. Then, pipes feed the propellants into an engine, where they burn and push the rocket in the desired direction (for more detail, see Chapter Eleven: "The Technical Stuff"). Liquid propellant rockets fix many of the problems that solid propellant rockets have, but also introduce some of their own. Liquid rockets can be throttled (think again of your car accelerator pedal) and shut down at any time. Rockets that run on liquid fuel can be relatively light and easy to move around, only made heavy when their fuel and oxidizer are added. Adding those propellants bring to light some of the problems with liquid propulsion. Most of the liquids useful for rocketry are hazardous, either by being extremely cold or extremely toxic. Liquid propellant engines as they're currently designed and built by the majority of space efforts are very complex. Many of the engines push the absolute limits of materials and engineering to get the maximum power from an amount of propellant.

Liquid propellant rockets remained a theory until March 16th 1926. On that day, a professor from Massachusetts test-fired a small liquid rocket that flew for 2½ seconds and 56 meters. His funding came from the Smithsonian Institution, and the benefactor wanted to hail the accomplishment for the breakthrough that it was. The professor in question, Robert Goddard, was ridiculed in earlier efforts to publicize his theories and efforts. He asked that The Smithsonian keep his work secret. Later, having been driven from New England by a fire marshal after one of his rockets went astray, Dr. Goddard moved his efforts to New Mexico. There, with new funding from the Guggenheim foundation, he improved his designs and his rockets reached greater altitudes as time went on.

After initially falling from military interest, rockets made their way into popular culture in Europe after World War I. Practical demonstrations of rockets pushing cars, aircraft, and even people were the rage at shows across the continent. Rocket societies sprang up to try and turn the showmanship into something a little more substantial, and eventually the German Society for Space Travel (abbreviated in German as the VFR) caught the eye of the military. As the Nazi party came to power in the early 1930s, the excitement surrounding rocketry faded away. National policy made it illegal to discuss rockets. The members of the VFR, including one powerful member Werner Von Braun, started developing the rockets that would shock the world in 1944 as Germany rained its "Vengeance Weapons" on London towards the end of World War II.

After WWII, the allied powers raced to snatch up information and personnel who contributed to the Nazi war effort. German rocket scientists were high on the list. Werner Von Braun and a group of his close advisers were brought to the United States and jump-started America's missile development program. Von Braun went to work for the US Army, building missiles that could carry nuclear warheads for relatively short distances.

Meanwhile, the United States was growing increasingly anxious about the capabilities and intentions of it post-war rival, the Soviet Union. The strict secrecy practiced by the USSR made it extremely difficult to gather information about the nation. The United States tried using advanced spyplanes to peer behind the Iron Curtain, but was frustrated by increasingly accurate anti-aircraft missiles, culminating in the shootdown of Francis Gary Powers in 1960. Some believed that an orbiting satellite could provide the desired intelligence information at a much lower risk, and the Corona satellite program was started. The Corona program was designed to circle the Earth several times, taking pictures of the Soviet Union. These pictures would be recorded on film and returned to Earth for developing and interpretation[29].

Along those lines, the increased accuracy of anti-aircraft missiles made military planners nervous as to whether or not large fleets of bombers carrying atomic weapons could reliably reach their targets in the Soviet Union. Looking for alternatives, they considered large missiles (called intercontinental ballistic missiles or ICBMs) that were capable of shipping a hydrogen bomb outside the Earth's atmosphere and accurately delivering its payload in "30 minutes or less, guaranteed"[30] on the enemy. These rockets were huge by existing standards and required meticulous construction and care while being readied for flight. Once readied, they were designed to sit for long periods of time with little maintenance, awaiting the moment that no one hoped would come: when they were rapidly fueled and launched in anger. This military mindset would play a critical role in all space activities for years to come, and it had its roots before the idea of a space race took over the American consciousness.

Tom Hill

A Quick Start: 1957 – 1968

On October 4, 1957, the Soviet Union astounded the world by announcing their crowning technological triumph. Using technology designed and built within their borders, utilizing German technology and scientists just as the United States did, the nation launched the world's first artificial satellite, called Sputnik 1, which is Russian for "traveler." Both superpowers hinted that they'd launch a satellite in 1957, in celebration of the International Geophysical Year, but the United States planned and fully expected to launch first[31]. This Soviet beachball-sized spacecraft circled the Earth every 90 minutes, broadcasting a "chirp" signal that any ham radio operator in the world could pick up. The satellite launch itself was not a threat to national security, but the capabilities that Soviet rocketry required to accomplish the launch were a threat.

This event caused the United States to do some much-needed soul searching. Panicked questions were asked like "Are they going to bomb us tomorrow?" but more important questions like "Is our education system geared to produce competitive scientists and engineers?" carried the day. People[32] who lived through the period described the radical changes that took place, adding more mathematics and science to their academic calendar. The furor only increased when a United States rocket and satellite, called Vanguard, rose a few feet off the launch pad and fell to the ground in a huge fireball on December 6, 1957. As a public response to this event, President Eisenhower created the National Aeronautics and Space Administration, or NASA. NASA's charter was to consolidate most space activities (the consolidated activities turned out to be scientific research and crewed spaceflight) under one agency. It wasn't until January 1958 that America successfully launched its first satellite, Explorer 1, under an Army program using one of Werner Von Braun's rockets.

The above-mentioned Corona program first launched in February 1959, but didn't return useful images until its fourteenth flight. This fact states something amazing about the difference between military and civil space: national priority (even, or perhaps especially, when that priority is classified) can keep funding available, even after failures. Of course, the fact that Corona was designed to provide data that was unavailable any other way was a big help. From that time until 1972, the program provided a consistent stream of imagery to intelligence agencies. After 1972, satellites whose capabilities remain classified replaced Corona.

Something that goes almost unnoticed during all this early activity is the efforts of a group of hobbyists that would later be known as AmSat. This group was interested in using space to work with ham radio. In December 1961[33], a member of this organization, who happened to work at one of America's launch sites, convinced the people launching a Corona satellite to use a home-built satellite as part of the required ballast weight on board the launch vehicle. This satellite, known as OSCAR I, broadcast "Hello" in morse code for the entire ham radio world to hear. AmSat became the third organization from Earth to have a successful payload in space. The powerful point to be made here is that while the mystique of space was being built up as something only professional, government-run organizations can participate in, a group of people, working at home, built their own satellite!

The first American ICBM, called Atlas, flew in December 1957. For comparison, the Soviet Union launched their first ICBM in May of the same year. The first two Atlas test flights were unsuccessful, but the third pushed its payload (for the test, just a dead weight, not a nuclear device) as required and delivered it to a target in the Atlantic Ocean. This booster was large enough to deliver the bomb to target, and it was also large enough to push a smaller payload into Earth orbit. Either job required the same meticulous preparation, however, and this played into the procedures of future rocketeers for decades.

A second round of soul-searching took place in 1961, when the Soviet Union launched Yuri Gagarin to complete one orbit of Earth on April 12th. US efforts were again called into question, since their first crewed flight wasn't scheduled until May, and it only involved launching a craft a few score miles away from the launch site. The newly inaugurated Kennedy Administration scrambled to find a way to turn this political tragedy into something positive. They fell upon an interesting idea, in that space spectaculars were a way to demonstrate superiority without direct warfare. The devices used to get to space were derivatives of weaponry, to be sure, but the warfare subtext would likely be lost in the celebration of a new achievement. There are also stories that Vice President Lyndon Johnson pressed for space activity as a way to industrialize southern states. In consultation with NASA, the administration made the decision to set the goal: "before this decade is out, of landing a man on the moon and returning him safely to the Earth." This goal was considered far enough ahead to allow the US to catch up and surpass the Soviets in the space race, and many people at NASA believed that the Soviet Union was not up to the challenge. NASA's reasoning was that going into orbit requires precise guidance for several minutes, while traveling to the moon required repeated precision guidance, especially in the return (arguably the most important) phase.

So, after all the teeth gnashing, President Kennedy formally committed the nation to the goal in a special session of Congress on May 25, 1961. While the stated deadline was the end of the decade, quietly NASA was asked to try and make the event take place in 1967, when it was feasible that President Kennedy would still be in office. I find it interesting to note that this speech was about much more than space travel and the moon landing. President Kennedy called for huge increases in federal spending elsewhere as well. This included a rather hefty tax increase, part of the details that appear to be lost in the glow of past achievements.

The race was on. NASA had already started preliminary work on a space capsule (to that point unnamed, although it would eventually be known as *Apollo*) that could orbit the Earth, orbit the moon, or, when combined with some additional huge components, land on the lunar surface. They also had the Mercury spacecraft, a one-man capsule that lofted Alan Sheppard and Gus Grissom to their heights before dropping them into the Atlantic, and would later carry John Glenn and most of the rest of the "Mercury 7" into orbit for flights that eventually lasted for more than a day. The problem arose in that Mercury was relatively primitive in comparison to the planned spacecraft, and current flight schedules had a multi-year gap between when Mercury would stop flying and the new craft would start. What to do?

An answer came in a proposed program called Mercury Mark II. In a bit of a coup, this new program came about as a contract extension to the original Mercury purchase, and McDonnell Aircraft Corporation was asked to improve on the original design and make a capsule that held two people instead of one. The Air Force, continuing its development of more powerful and more flexible ICBMs, provided a launch vehicle called Titan II for the spacecraft, and the Gemini program was born. Gemini would allow NASA to experiment with procedures, technologies, and studies critical to the eventual moon effort. Changing orbits, long duration space flight, rendezvous (approaching an object that is already in orbit), docking (joining with that object and forming essentially one spacecraft), and Extravehicular Activity (EVA, or going outside the spacecraft in a space suit) would all play major roles in landing on the moon.

The Gemini program flew from 1965-1966, and showed that some perceived problems would be relatively simple (rendezvous and docking, while not easy, was never very difficult), while other things thought to be simple turned out to be big problems (it took several tries and a whole new approach to training to get EVA to work, and fuel cells, devices that turn hydrogen and oxygen into electricity and drinkable water, spent years in the development phase). Gemini also took advantage of its time to do some amazing things that were then forgotten for years by "mainstream" space. One example is artificial gravity generation using a tether or rope between two objects. Pete Conrad and Dick Gordon in *Gemini* 11 tied a tether to their target vehicle after docking with it, then backed away and started both spacecraft spinning around their common center of mass. By doing this, they created a feeling of gravity much the way a child spinning a bucket of water around can keep the water in the bucket[34].

During this time period, a number of uncrewed space missions rapidly opened our eyes to the universe around our planet. Probes to the moon, Venus, and Mars started things off, and plans for missions to go even further were in the works. Usually, a series of missions would include a flyby, where a spacecraft made a fleeting pass of the planet, taking a few pictures and beaming them back to Earth. Next, an orbiter would fly and stay for a while. Eventually, landers were designed to settle down onto the planet.

After the triumph of Gemini, the now-named Apollo program stumbled badly when three astronauts, Gus Grissom, Ed White, and Roger Chaffee, were killed in a ground test aboard their *Apollo* 1 spacecraft in January 1967. Their deaths forced the space agency to reexamine its priorities and methods that it was using to achieve President Kennedy's goal. A top-to-bottom review of procedures and hardware took place, and caused major changes in how things were done. Many who worked on the Apollo program believe that, without the *Apollo* 1 fire, something else, worse, would have happened and prevented the lunar goal from being reached.

After a pause lasting over a year, Apollo roared back with a series of missions that built upon each other to beat the path to the moon in a matter of months. *Apollo* 7 put a crewed capsule in space orbiting the Earth. *Apollo* 8 took the same type of capsule and sent three men around the moon for Christmas, 1968. *Apollo* 9 checked out the much-delayed lunar module in Earth orbit, while *Apollo* 10 carried out a dramatic dressed rehearsal of a lunar landing, coming to within 10 miles of the lunar surface. The seeds of future disinterest were already sown, however.

Increasing social unrest at home and an escalating war overseas took attention away from space activities. When attention wanders from a task in government, the money isn't far behind, and NASA found out what wasn't widely known in the public sector, but a very large explosion dimmed the Soviet Union's hopes of reaching the moon when their moon rocket was lost while sitting on the launch pad. Then in 1967, the United Nations approved the space treaty, forbidding a nation from claiming a heavenly body. Some believe that the US ratified the treaty as a deliberate action to take the heat off the space race…if the Soviets were unable to legally claim the moon, then the Russians' landing on it wouldn't be such a big deal, would it?

Throughout this time period, while the manned missions gathered the big headlines, and the uncrewed missions to other planets came in a close second in hoopla, other events took place that probably had a bigger impact on our day-to-day life. Some avoided the public eye on purpose, through military secrecy. Others missed the spotlight because they were not showy enough to make good copy next to the headline-grabbing firsts.

The first weather satellite flew in 1960, called TIROS for Television Infrared Observation Satellite, the craft proved that images of clouds taken from space were useful. Initial plans were for a weather satellite that would serve both military and public needs, but continued differences of opinion on what the satellite needed to do between NASA and the Air Force caused the military agency to go on its own and adopt a low-cost option that met their requirements. Later refinements to weather satellites moved them to other orbits, providing a constant view of one side of the Earth.

Military personnel also looked to space for a universal solution to navigation. Local systems, near airports, allowed aircraft to know where to find safe harbor. Also well-traveled routes within the United States were well marked with radio beacons to guide pilots to their destination. The military required this type of service all over the world (at times when the "host" nation was not likely to provide it) and developed a series of satellites to answer that need. The first was the Transit system, which launched in late 1959.

Military planners had their own short-sightedness when it came to some programs, however. One military-man-in-space program, called Dyna-Soar (let that be a lesson to any future program directors out there: DO NOT name your vehicle after an extinct animal. It makes the jokes way to easy when things start looking bad at budget time) faced the axe because of rising costs. Meanwhile, the Manned Orbiting Laboratory, or MOL, was developed to allow two men to fly into orbit and take images of the Soviet Union. Unfortunately for MOL, having people on board a spacecraft does bad things for that craft's picture-taking capability. As these realities set in along with a rising budget, the program was scrapped, but not before building a beautiful launch site at Vandenberg Air Force Base.

While these first two examples may seem a little obscure, the next is as close as your television. Communications were an early use envisioned for satellites, described by Sir Arthur C. Clarke as early as 1945. While one of the masters of science fiction imagined huge, crewed switching stations in the sky, the communications satellite industry has grown into the first space application that can support itself through selling its services. Modern-day, non-fiction communications satellites are relatively small spacecraft circling the globe. This entire endeavor started in 1963, with the signing of the Space Communications Act. The act created an organization called COMSAT to investigate and make use of satellite communications.

Triumph and Waning Interest: July 1969 – 1976

On July 20th, 1969, the world paused for a moment to watch an event take place that was like no other in human memory. In any nation where television technology was possible, people gathered from miles around to watch a human being put his footprints on another heavenly body. Neil Armstrong was that man, and after his "giant leap," Buzz Aldrin (he's had his name legally changed to Buzz, by the way) joined him on the outside of their spacecraft. For two hours these men walked around the surface of the moon and gathered samples of the lunar rocks. They then returned safely.

Unfortunately for public interest in the space program, the last two sentences of the previous paragraph pretty much describe what most of the other Apollo moon missions accomplished. There were changes, of course, in that the *Apollo* 12 astronauts visited a spacecraft named Surveyor that the United States placed on the lunar surface 3 years before the crewed mission. *Apollo* 15 took a lunar rover along with them, allowing them to explore a wide expanse of sites around their lander. The *Apollo* 17 crewmembers spent nearly an entire day walking on the surface and took along the first (and so far only) trained scientist to walk on the surface of the moon. While these differences were huge in the eyes of NASA workers, astronauts, and those who built the hardware, the public basically said "so what?"

The only mission that caught the public's imagination, for all the wrong reasons, was *Apollo* 13. The star-crossed voyage of Jim Lovell, Jack Swigert, and Fred Haise held people glued to televisions around the world again, just to see if the crew could make it home. In the end, a lot of ingenuity and some good luck held things together long enough to bring the boys home. One thing to note here, because the *Columbia* disaster is consistently compared to *Apollo* 13, is that the *Apollo* 13 vehicle was essentially sound, and no further launches were required to bring the crew back safely. While the effort necessary to bring the crew back was impressive, it really meant nothing more than some late nights on the part of ground controllers. Rescue plans for

Columbia, in contrast, required the quick launch of a space shuttle. This would be possible, but extremely risky. More on that later.

During the autumn years of the Apollo program, a lot of study money was thrown around to talk about the next step in space travel. Study money is perceived as a good thing in government space programs because it keeps people employed, and studies don't cost a lot. In the days before Microsoft Powerpoint (the critical link in any organization today) a study kept engineers, artists, and managers busy producing presentations of what was possible. The concepts ranged from the relatively simple, such as landing a spacecraft on the far side of the moon, as depicted in James Michener's *Space*, through the more difficult task of building a large space station in Earth orbit. Pie-in-the-sky plans called for everything—reusable spacecraft, space stations, lunar bases, and an expedition to Mars.

As the President hinted in his conversation with the *Apollo* 17 crew on their return to Earth, 1972 saw the last lunar landing of the 20th century. The United States' leadership moved on to other priorities, and a stagnating economy and difficult war overseas pushed space travel low on the agenda. A task group set up to determine what the next step should be proposed a reusable spacecraft as NASA's next project. Once this spacecraft was fully operational, traveling between the surface and orbit with the advertised regularity of a passenger jet, NASA would have a low-cost ability to build their space station, and all their other projects.

First, some leftover *Apollo* hardware was put to use. This additional hardware was assembled to form a space station known as Skylab. Launched aboard a single Saturn V moon rocket, the space station went into Earth orbit with some initial problems including low power and extremely high on-board temperatures. With some ingenuity and a couple quick-fix solutions (along with, arguably, the most useful spacewalk in history to restore power to the craft) Skylab supported 3 crews of astronauts for a total of nearly 6 months total time on orbit. At the time, US astronauts held the endurance record for people living in space.

After Skylab, leaders of the two spacefaring nations decided to showcase their new policies of peaceful cooperation by holding a joint mission. *Apollo-Soyuz* flew in 1975, where a US Apollo craft docked with a Russian *Soyuz* vehicle. This first experience with combined training would serve as a challenge inspiring future cooperative events.

The Soviet Union, after another failure of their moon rocket, claimed publicly that the moon was never their goal, and poured resources into a manned space station program. A series of stations named Salyut (translates to "Salute") demonstrated increasing time on orbit for crews. This program was not without problems, however, as the first Salyut crew, after flying a nearly flawless mission, was killed upon their return to Earth when a cabin valve opened and allowed their atmosphere to escape into space[35].

Uncrewed spacecraft continued to press the boundaries of the possible during this time. While the Mariner program continued to explore Mars, the Pioneer program sent two spacecraft to the outer planets, returning the first up-close information from Jupiter and Saturn. In one unique mission, the United States sent an uncrewed biology laboratory to the surface of Mars, claiming that it would answer the question of whether or not there was life. While some consider the results controversial[36], the majority of the scientific community agrees that the *Viking* landers did not find life in the surface soil of Mars. As scientists gathered more information about life on Earth, it became clear that life on Mars could not be completely ruled out through the *Viking* experiments, either. Also, towards the end of this timeframe, the United States launched a deeper survey of the outer planets. The twin spacecraft were called *Voyager* 1 and 2.

Meanwhile, in the background, hobbyists continued to develop their efforts. AmSat launched 3 home-built satellites during this timeframe, increasing their satellites' capabilities while continuing to build the little spacecraft on their kitchen tables. AmSat's new satellites were capable of storing messages for rebroadcast around the world. The communications satellite industry moved into the realm of treaty organizations and the amount of data beamed around the world (known as bandwidth) grew amazingly. In these early days of satellite communications, television signals and telephone calls were the primary data types.

During much of this time, Europe worked on a project that the US didn't consider very important at the time. The European Space Agency (ESA) decided to get into launching satellites. They built a rocket, called Ariane, and a launch site in South America to launch their rocket from. When this decision was made, America was convinced that they'd rule the satellite launch market with their space shuttle, still advertised to be the cheapest ride into orbit, with the benefit of being able to repair satellites as required and eventually demonstrated on shuttle missions. When Ariane first launched in 1979, and then flew in more powerful versions throughout the '80s, they were poised to take advantage of the upcoming hard times for the US space program.

The manned (at the time, no US women had flown yet) United States space agency poured all its efforts into its next project, The Space Shuttle (officially known as the Space Transportation System or STS, that's why space shuttle flights are all labeled STS-XXX). The space shuttle was sold as a method of decreasing the cost of sending payloads into low Earth orbit. By extension, anything else that needed to move beyond low Earth orbit would cost less to move, too. Old charts presented during the early shuttle program advertised shuttle launches to cost $10.5 million[37], an amount that was estimated low by between 50 and 100X, but more on that later. In a major coup, NASA, to justify the cost of the space shuttle, had it written IN LAW that all space payloads for the United States were to fly on the space shuttle, including military payloads. In 1976, the first space shuttle, *Enterprise*, was rolled out of the factory. Enterprise would never fly (I'm curious if the Star Trek fans who wrote to NASA by the thousands and asked that the first shuttle be named after the craft from that series were aware), but the event was hailed as the dawn of a new age of space exploration. It was a new beginning, but of what?

A Space Truck...to Where? Reality Strikes 1977 – 1986

From 1976-1981, no American flew into space. The biggest news related to crewed space flight from the Western Hemisphere was the fall of Skylab. The space workshop was left in orbit at the end of its mission, and expansion of the Earth's atmosphere pulled it back to Earth. Amid much media hype (for the time—remember that there were only three major networks back then) Skylab burned up in the atmosphere and scattered pieces across parts of Australia.

Space shuttle development plodded on, and NASA discovered (again) what Norm Augustine described as "The High Cost of Buying." It turns out that when you're doing something that's never been done before, in most cases it will cost more than you expected. This did not deter the space agency, however; they had a sure fire plan. The original plans for buying space shuttles called for 8 orbiters—4 for civilian use and 4 for military use. As costs rose beyond a certain threshold, they would delete one of the ordered shuttles. Unfortunately, buying space shuttles is not like buying balloons. If you originally want to buy 8 balloons, but don't have the money, you can only purchase 7 and you save 1/8 of your original cost. Purchasing space shuttles works a little differently in that cutting a purchase from 8 to 7 likely cuts about 1/20 of your cost, because by the time a construction crew is building its 8th space shuttle, the people are very good at their jobs. In the end, 5 space-ready orbiters (remember, not counting *Enterprise*, but including the replacement for *Challenger* named *Endeavour*) were purchased at a cost that was still greater than originally budgeted.

A series of approach and landing tests made by *Enterprise* gave some people a taste of space event news coverage. In these short missions, the shuttle flew on the back of a 747 aircraft, and was then released to fly back to land on a runway.

The first space shuttle launch took place in April of 1981, two years later than expected by most estimates. The first four flights took place with a crew of two, and each astronaut had an ejection seat to throw them away from the craft in case of a serious problem in the first moments of flight. These missions allowed the craft to flex its muscles, and put a huge chain of events required to get the vehicle back into the air into motion between flights. The first missions are my first memory of live space flight. I even did an art project related to the shuttle's first flight, and got an A.

The next 20 space shuttle missions went relatively smoothly, with an occasional errant satellite catching the media's interest (after the first handful of missions, the only newsworthy events were liftoff, landing, and maybe a significant mission event—unless there was a problem, then all ears tuned quickly). The US space agency seemed to be catching onto a pace, with a space station proposed by President Reagan as their next project, and a series of shuttle missions were scheduled to launch and repair satellites, as well as launch space probes and carry out science experiments while building the promised space station. The Air Force was gearing up for its first mission launched from Vandenberg AFB (see the technical details chapter to find out why this is a big deal) and the crew included the soon-to-be secretary of the Air Force, Pete Aldridge. All these plans changed on a cold January morning in 1986.

One way NASA considered catching the interest of young children was to put a teacher in space. A call for volunteers went out, and thousands submitted the required paperwork to become the first educator to teach from space. After a lengthy evaluation process, Christa McAuliffe was chosen to fly on board the Space Shuttle *Challenger*, using the new, cryptic notation of STS-51L for the flight. 51L denoted the year the mission was originally scheduled to launch ('85, in this case) the launch site (1 meant Florida, 2 meant California) and L showed the order in which the mission was supposed to take place. In my opinion, the flight numbering scheme was changed because launches were happening out of a straight numerical order, and people in charge didn't want it to be obvious that plans were switching too much.

In a media frenzy that any Hollywood producer would be proud of, Christa made the rounds of the talk shows after her training peaked and before the flight, and took her role of the first "everywoman" to go into space with style. As the seven shuttle crew members made their way to the launch pad, smiling and waving to the crowds, everything looked ready for another excellent shuttle mission.

Everything was not ready for an excellent mission, and there were people on the ground dreading the moment when the engines would fire. Engineers at the Morton Thiokol Corporation knew that there'd been a series of missions where the solid rocket boosters (the two white, pencil-shaped objects on either side of the big brown tank in the middle of the shuttle at launch) almost failed. Such a failure would have caused hot gasses to shoot out the side of the rockets and caused the space shuttle to explode. When the shuttle launched on a cold day, these boosters came closer to failing, and the scheduled launch day for *Challenger*, January 28, 1986, was the coldest day a shuttle had ever been launched on. The engineers took their concerns to their managers, who took them to NASA, climaxing in a meeting the night before *Challenger* was scheduled to take flight. Thiokol's statement was essentially, "We're not sure we should launch." After looking at the data, NASA managers essentially said, "According to this data, we should never fly. Are you saying that you built bad hardware?" This put the engineers and managers at Morton Thiokol in a very bad situation. If they agreed that their hardware was bad, they'd possibly lose a lot of business, but if they said the shuttle was good to launch, seven people might die. Unfortunately, the climate created at the meeting, along with some very wishful thinking, made the NASA and Thiokol managers approve the shuttle for flight.

That decision created one of the defining moments of the 20th century. When the *Challenger* lifted off, a small puff of smoke appeared along the side of one of the right solid rocket boosters. As the craft flew out over the Atlantic Ocean, a plume of superheated gas shot out the side of the SRB, and eventually started to cut into the brown external tank. Soon, the tank started to leak, and soon after that the solid rocket booster cut partially loose from the tank. Seconds later, the *Challenger* was engulfed in a fireball that awed and confused the people watching. Seven lives and one-fourth of the United States' space fleet were gone.

Three other US rockets exploded in the next couple months, and the United States was without a way to get anything into orbit. While the loss of those rockets impacted several agencies around the country, the loss of the *Challenger* shocked the nation. On a personal note, one of those explosions happened on the West Coast, at Vandenberg Air Force Base, where I served at one point in my Air Force career. The hillside where the accident happened is still littered with pieces of the rocket that weren't useful to the investigation. On a whim, one day I walked out and collected a souvenir.

A presidential commission, headed by William Rogers, formed to get to the bottom of what happened on January 28th. They looked into the design of the spacecraft, the faulty decision-making process at NASA and the companies that built the shuttle that lead to the dreadful decision to launch the crew on their mission, and the culture that drove NASA to try and do more and more with their spacecraft. All the factors played a role in the disaster, and all of them had to be fixed, or at least faced, before the shuttle could fly again.

REGAINING FOOTING, BUT STANDING ON WHAT? 1987 – 1995

As the Rogers commission completed its report, NASA worked to change the way it did business. This required both hardware (improvements to the shuttle to prevent another disaster) and organizational (new decision making processes to prevent disastrous decisions like launching *Challenger* that day in January) changes to be completed.

The most critical fix was to the shuttle hardware itself. Investigation showed that a field joint, connecting parts of the solid rocket booster (SRB) failed, causing the explosion that took the lives of the *Challenger* crew. The solid rocket boosters are built in the factory in pieces and then shipped to the Kennedy Space Center (KSC) for final assembly. Because this final assembly doesn't take place in the factory, the connections between the pieces are known as field joints. The original field joints (up until January of 1986) were made of two pieces of material that fit very closely together. In between these two pieces were seals known as o-rings. The o-rings acted like the rubber lip around the inside of a metal lid to a glass jar, sealing the joint and preventing hot gasses within the solid rocket booster from escaping. These gasses, once loose from the SRB, acted like a blowtorch and cut through the large external tank holding fuel for the *Challenger* on its ride to orbit.

As discovered in investigation, this joint worked much better when it was warm than when it was cold. Previous launches showed some leakage of the hot gasses (called "blow-by") on several missions, but much more blow-by when the shuttle was launched on cold days. Despite all the tests that demonstrated this conclusively, the biggest public relations moment came when board member physicist Richard Feynman took a piece of the o-ring material and showed it to the press, emphasizing how pliable it was. After that demonstration, he dipped it in ice water for a moment, and showed how solid the ring had become, cutting its ability to protect the crew.

To fix this problem, NASA and its contractors added an additional lip of material to hold the field joint in place. They also added a third o-ring to make the seal solid. To make absolutely sure that the system was safer (you'll note that I didn't say "absolutely sure that the system was safe"—that was done on purpose) than its predecessor, joint heaters were added to the system.

NASA took this two-and-a-half year time out to address some other hardware problems as well. New pumps for the space shuttle main engines were ordered, promising to require less maintenance than the ones originally put on the spaceplane. An emergency exit was added to the spacecraft, in the form of a long pole that spit out the side of the shuttle when it was flying straight and level. The astronauts would attach themselves to this pole and bail out over the ocean to be picked up in their own personal rafts. This device, described as "phase 1" of the crew escape system, rapidly turned into the only option a crew had to get out of the orbiter. Unfortunately, escape using this system required the craft to be stable and flying relatively slowly.

Perhaps the more difficult changes had to take place within the agency itself. NASA looked in the mirror (or more correctly, had a mirror held in front of it) and saw that they'd promised a routine access to space for the nation, but that their spaceplane was shaping up to be much more work than promised. Early flight rate estimates had space shuttles each flying once a week. Given that the initial fleet had four orbiters, that could mean four flights a week, leading to a dizzying two hundred flights a year. Published works on the shuttle program never predicted that kind of flight rate, but did say that year 6 of shuttle operations would see 49 flights[38]. A major problem with this idea was that NASA was not made to run a recurring operation.

When NASA was founded in 1958, its charter was not to run an interplanetary bus service, but to focus United States' efforts in space. It's predecessor, NACA, was a much smaller agency, and applied its limited budget to researching technology that would help advance aviation. NASA, on the other hand, actually conducted crewed and uncrewed scientific space operations for the nation. When such missions were brand new and cutting edge (read as: exciting), NASA was in its prime. The Mercury, Gemini, and Apollo missions were constantly trying new things in their efforts to explore the moon. By definition, then, the space shuttle, billed as a routine way of accessing space, was not something NASA was built for. While the argument holds true that no shuttle launch can be considered routine, repetition of the same operation, no matter how complex, drives the excitement out of each individual mission. It also brings up new managerial challenges, which the space agency was not ready to face.

Transcripts from meetings held before the *Challenger* launch have the definite tone that managers wanted their engineers to prove to them that there was a problem that would prevent the launch. Engineers wanted to delay the launch until a warmer day, because they had experience with blow-by in previous launches on cold days. Launch managers brought up the fact that blow-by took place in warm-day launches as well...did this mean that the entire joint design was faulty? As mentioned before, this question put engineers (and likely more importantly, their direct supervisors) in a difficult situation. If they said that the joint design was faulty, and called for a redesign of the entire system, it was possible that the space shuttle would be grounded for years. This sort of thing doesn't sit well with a government agency, or any organization for that matter. So, while engineers refrained from saying that the joint itself was faulty, they implored management to wait a few days for a warmer day to launch. The engineers' supervisor agreed with the assessment, and recommend against launch, until, in an infamous quote, launch managers asked the engineers' supervisor to "take off your engineer hat and put on your management hat." The launch was approved for the fateful day.

Several books have been written about the decision to launch *Challenger*. To me, the bottom line is that a group of people who built something were put in the position of stating that their hardware was good to fly when they had direct evidence that it may not be. In an interview, one of the engineers in question said that he watched the *Challenger* launch, expecting the shuttle to explode on the launch pad. When it cleared the tower, his supervisor squeezed his shoulder and said, "Looks like we dodged a bullet." Of course, moments later, both knew that the thought was wrong.

While there will always be engineers wringing their hands over some problem they're convinced can bring the end to a space flight, when that engineer has data supporting their fears, they should be listened to. When a system starts flying enough to create statistically significant numbers (remember that the entire manned Apollo program to the moon only consisted of 11 flights), items of concern, such as blow-by, on more than one flight under different conditions is a cause for alarm.

In response to this NASA adopted a safety policy that put the burden of proof on the managers. Instead of management making the engineers prove that an action is unsafe (this was essentially the tone of the meeting before the *Challenger* flight), it became everyone's responsibility to prove that the vehicle was safe. In theory, this approach means that *Challenger* would not have flown, although there are some indications that this thought process eroded over the years between *Challenger* and *Columbia*.

National policy also changed. The earlier decree that all national payloads would fly on the space shuttle was rescinded. This cleared the way for the US military to purchase more expendable launch vehicles for launching their many satellites, but it also cut into the number of flights NASA had planned for the shuttle. Cutting further into the manifest, NASA was forbidden to fly commercial payloads on board the space shuttle, and President Reagan stated that the shuttle was to be used only in cases where human assistance was required.

In the fall of 1988, the space shuttle *Discovery* rocketed away from the Florida coastline in the first post-*Challenger* flight. The flight was successful, and got the space program back on track, but on a track to where? Only a few military missions flew, in final closeout before the missions moved to Air Force expendable launch vehicles. With no commercial customers, only US civil spacecraft could fly on board, and there weren't enough of them to make a case for the entire shuttle program. Occasional science-devoted missions provided a more obvious (but not necessarily more correct) reason for humans to be in space, but not many of those were on the books, either.

Other rockets that failed in 1986 returned to flight much quicker, because they did not have crews to worry about. Through increasingly commercial operations, these expendable launch vehicles took on Europe and other launch providers in the expanding business of telecommunications satellite launch. The Department of Defense switched to these rockets exclusively by the end of this period, launching weather, navigation, communications, surveillance, and other types of satellites on board.

The United States space agency, however, was in a quandary. Most of the payloads for the shuttle were taken away, and the space station, long-planned and much-researched but not in existence yet, was years off. The space station, once construction started, would provide the ultimate reason to have people in space…to build the space station, but humanity's outpost in space faced its own hurdles to become a reality.

In February 1993, the administrator of NASA, Dan Goldin, received a call from the White House. As it turned out, NASA was in line for another budget cut for the coming year, and there was no way that the space station could ever be built as envisioned with the new funding levels. In the midst of a mad scramble to re-plan the station, an idea came to light. Why not ask the Russians to participate? The early nineties was a rough time for the Russians, with economic troubles catching up to the cold-war Soviet Union and splitting the empire. There were a lot of scientists that now had no jobs, and idle scientists make people nervous. They could go out and offer their services to undesirable nations, and eventually cause great harm. If Russia was brought into the space station program, those scientists would be funded and kept busy, and therefore be less likely to freelance for other jobs.

So, the deal was done, and amid much splendor and promise of cost-cutting, Russia joined the aptly, but unimaginatively, named International Space Station. Russia would develop and launch the first part of the station, and the US would fund other portions while Russia built them. Delays started soon after the new program was announced, and soon entire websites, booming with the internet economy, would chronicle the delays as they happened.

Shuttle operations continued, though, awaiting a destination, and in a stirring speech that President Bush gave in 1989 on the steps of the Air and Space Museum on the 20th anniversary of the *Apollo* 11 moon landing, he gave them one. Attempting to resurrect the excitement of space travel in a new generation of Americans, President Bush challenged the United States to complete the space station before the 30th anniversary of *Apollo* 11, return to the moon by the 40th, and land on Mars before the 50th. NASA immediately went to work designing a system that would meet this amazing goal, and after 90 days (leading to its name, "The Ninety-Day Report"), came back with an estimate of the cost: $450 billion. The silence was deafening. The perception was that there was no way the US government would spend that kind of money on space. The plan never received serious support. It should be noted, however, that the timeline (30 years) meant that a budget of $15B a year would accomplish the goal. The United States' defense budget is many times that value, and, to be fair, the federal government's expenditure on education, as requested by the President, in 2005 was almost four times that number.

The rise and fall of this mission to Mars idea got an engineer in Denver thinking about ways to make the trip less expensive. Robert Zubrin, with his friend David Baker, brought together a lot of ideas, some of which were proposed before but never in a group, to form the Mars Direct mission plan. Mars Direct turns many ideas that were used to create the Ninety-Day Report on their head. In the plan, the space station is not necessary, and neither is a moon base. Small crews travel to Mars and make most of what they need from the resources there. And the cost is 10% or less of the budget listed in the 90-day report. This is done by launching the Earth Return Vehicle, or ERV, to Mars first, with a chemical plant and chemicals. Using Mars' atmosphere, the plant and the chemicals carried on board, the ERV fills itself with rocket fuel, oxygen, and water. Once the ERV is confirmed to be filled with the necessary supplies, the crew is launched from Earth in a separate vehicle, called the habitat. The crew lands near the ERV and explores Mars' surface for over a year. Then they climb aboard the ERV and fly back to Earth.

The Soviet Union, as named at the beginning of this period, continued largely with business as usual during this period, with one exception. Where previous eras saw the USSR launching space stations individually, to be inhabited for varying periods of times by crews, this era saw a focus on one space station. The Mir (Russian for "peace"…I remember being quite cynical at the name as a youngster when I heard the translation) station was designed to be modular. The primary module was launched in 1986, and follow-on modules were launched up to it to dock and increase the living space aboard. Crews eventually rotated through the station, keeping it constantly occupied. In the process, they set increasing long-duration stay records. As time wore on, the leadership of the USSR saw value in inviting travelers from other nations within the Soviet Bloc to fly to Mir. In an ultimate twist of fate, eventually even the United States sent its own astronauts to board the Mir station, in what was called Phase 1 of the International Space Station program. A total of 7 Americans flew aboard Mir, and the images taken of the space shuttle docked to the Russian station grounded the ideas of former cold war rivals working together in truth.

Telecommunications services via geostationary-orbiting satellites blossomed during this time period. As data rates between points on Earth increased, satellites grew in power and complexity to meet the growing demand. Launch to geosynchronous orbit became so commonplace that new ideas came about for using the location. These ideas included direct broadcast satellite TV and satellite radio broadcast to cars. These new startups had the advantage of building upon a well-known "location" in space.

There were also isolated efforts in the area of new commercial satellites (meaning other than geosynchronous) and launchers. Connestoga, Iridium, and Teledesic are some examples of the trade names involved.

In a "way out there" move, the US military, as part of its Strategic Defense Initiative (SDI), built a small, cone-shaped demonstrator to show that rocket travel could be made simpler (and cheaper) than it was at the time. The DC-X, as it was called, could not travel into space, and only ever went a few thousand feet into the sky. Here, the amazing thing was the number of times it flew, the short turnaround times between flights, and the small ground crew required to keep the vehicle going. This craft had all the makings of a full-fledged rocket: engines, a light structure, and a sophisticated guidance package, the things that other groups say require hundreds of people and weeks of preparation. When the DC-X flew, however, 12 people could prepare it for flight and in one case it flew twice in one day. Inspired (or embarrassed) by this accomplishment, NASA stated that the time had come to develop single-stage to orbit vehicles, and requested that contractors bid for building it.

In another surprise, the US Department of Defense sent the first spacecraft to the moon since 1972. The Clementine mission, launched in 1994, was designed to demonstrate sensors for missile defense weapons. Testing sensors like that near Earth was frowned upon, so the research team decided to use the sensors to look at the moon and eventually an asteroid. This mission was developed, built and launched on a timeline that made NASA's process look glacial. The mission was largely a success, returning amazing images of the moon and providing tantalizing evidence that there was ice at its south pole. After leaving the moon, a software error caused the craft to fire its rockets until there was no more fuel left, and from then on the mission was abandoned. The message got through that small, focused missions could be launched quickly and provide a science return, though, and NASA announced that a new round of missions would take place designed to be faster, better, and cheaper than the norm.

In the scheme of things, all of this news was rather tame stuff. Nothing that was going on in space grabbed human imagination and shook it hard…that is, until a day in August 1996.

A Bombshell, and Business/Cooperation is (is not) the Answer, 1996 – 2002

The late 1990s was an amazing time for the world economy. Globalization and the seemingly magical powers of the internet spurred economic growth that for years before was described as impossible. With the rapid spread of technology through the industrialized world, fueled by unprecedented investment by venture capitalists to start new businesses, many companies turned to space as the next logical extension of their communications needs. For me, though, the bigger event took place at a press conference in August of 1996.

At this press conference, a group of researchers in Houston, Texas, announced that their study of a meteorite found in the Alan Hills area of Antarctica provided an unprecedented mix of chemicals. The scientific community had long since accepted the fact that this meteorite was from Mars, because the measurement of certain chemicals within the rock matched samples taken from a spacecraft called *Viking* 1 that took samples of Mars in the 1970s. The press conference was held because within a very small portion of the rock, there were chemical traces that, when all taken together, signaled the possibility that life once existed on Mars. The argument went that each of the chemicals on their own was not an indicator of ancient life, but when they were all found together, life was the simplest answer that could produce them all.

This announcement was billed as the start of the scientific process. The researchers were announcing their initial findings, and wanted other scientists to check their work to see if they missed anything. Over the years, I've read several papers both for and against ancient life in the Alan Hills meteorite. I don't have the experience or knowledge to have an important opinion on the topic, though it seems each side focuses on its strongest arguments, ignoring other portions of the investigation where their data is thinner. So, the scientific argument turns into nothing more than a shouting match on CNN's *Crossfire*, only through scientific papers instead of live television. Time will tell what the final consensus falls to.

When the potential fossilized life in the Alan Hills meteorite was discovered, two Mars missions were already far into construction. The *Mars Pathfinder* lander and the *Mars Global Surveyor* were preparing to launch in December of that year. Part of the announcement from NASA stated that it was too late to change the mission focus for those missions, but that the Mars exploration budget would be increased. With the increase in funds, NASA would be able to send two spacecraft to the Red Planet at each launch opportunity. Over time, these spacecraft would become more sophisticated, and eventually be able to look for more information related to the Alan Hills meteorite. The results of this announcement were mixed at best.

Of course, *Mars Pathfinder* went on to become the first internet event, beaming back images of the surface of Mars, along with pictures of its little rover, Sojourner, as it drove around on the surface. *Mars Global Surveyor* entered orbit at about the same time, and after some initial difficulties in reaching its final orbit, started returning unprecedented pictures and topographical data from Mars. It's still functioning today, continuing to send back images and relay data from the rovers on the planet's surface. Other missions were less successful.

The *Mars Climate Orbiter*, and *Mars Polar Lander* spacecraft, launched in 1999, came to a bad end on Mars. This is described in more detail later, but the orbiter ran into troubles because the builders of the vehicle worked in

English units, while controllers expected data to be in metric. The lander, it's surmised, ran into budget problems that prevented it from being tested completely, and a flaw that would have been found with more testing caused it to impact Mars' surface much faster than it was meant to.

I was particularly close to the space launch business during much of this period, due to my wrapping up a tour in Turkey and returning to the United States for an assignment as a launch operations officer at Vandenberg AFB, CA. The biggest news during the timeframe was the fact that the number of commercial launches each year was rapidly approaching (and eventually beating) the number of government launches each year. This trend peaked in 1999, and eventually the numbers settled back down towards what the world was used to, for right or wrong, of government launch numbers beating those of their commercial counterparts.

The crush of commercial launches came about because of the booming economy of the late 1990s, along with the perceived importance of keeping in touch anywhere on Earth as communications systems grew more important to daily lives. To answer this call for increased communications (both telephone and digital) several corporations arose, all with some pretty radical ideas as to how to meet their customers' needs.

The most infamous of the lot is Iridium. Conceived in the early nineties, Iridium's goal was to provide one worldwide network for cellular communications using a satellite constellation of 77 satellites. This is the number of electrons in an Iridium atom, which is how the system got its name. Later, as the constellation developed, the number of satellites dropped below that magic number to 72, but once you have a good marketing scheme, it's not a good idea to change. Besides, who would buy a system named Hafnium (atomic number 72)? The Iridium system set new records in numbers of satellites purchased at once and speed in deployment. Up to 9 Iridium satellites could be launched at once, and they could be launched from Russia, the United States, or China. I personally witnessed many of the launches from Vandenberg, usually happening once a month. As their launches progressed and the constellation filled out, Iridium's stock price flew high because it looked like the system was going to work. There were some problems with the satellites, in that some failed completely soon after launch, and most of the others suffered some sort of component failure that left them one more component failure away from a satellite-ending problem. Nothing could stop the juggernaut, however, and in 1999, the system was unveiled.

Well, there were some problems. Some were technical, while others came through a misunderstanding of the market. I'll go with the technical ones first. The phones were, even by 1999 standards, huge! The company realized that to be competitive, an Iridium phone also had to be compatible with cell phone services, and in combination with the large antenna and batteries required for the satellite system, led to a portable phone much like a cross between the early cell phones of the late '80s and a World War II infantry phone. Another problem arose in that the signal couldn't penetrate a building. Though Iridium claims that this fact was made public early on, I don't remember hearing it until late in the deployment phase of their system. If you wanted to talk through the Iridium satellite system, you had to go outside, and even when you did go outside, the call didn't sound very good, because of a relatively low data rate.

On the marketing end, the company didn't take into account the fact that the market would change while they developed the complex satellite constellation required to support the system. Initially envisioned for use in developing countries where cell phone service was spotty, Iridium didn't plan for the fact that it's much easier for a developing country to place a cell phone tower in a town than install a land-line phone system. By the time Iridium got on line, many of the countries the company envisioned calls coming from had regular cell service. By using that local cell service, Iridium users could get around the biggest marketing misjudgement of all, the price. Iridium air time was to cost $3 a minute.

Iridium landed with a big, dull thud, and brought down a lot of dreams with it. Eventually, the assets were sold at auction for less than 10 cents on estimated dollar value. At last word, the US military is the primary customer for Iridium service, though the rates they're charged by the new owner are undisclosed.

Globalstar also worked to create a cell-tower-less cell phone system, but by using a simpler system design cost much less to implement. This translated to a much lower per-minute user cost, and kept the company more solvent.

The darlings of the time period were the direct-broadcast TV systems. Designed to use satellites similar to those already in use by communications companies, these services were to take on the cable companies by providing competition in premium television programming. Customers were required to mount a small antenna on their house and point it at a spot in the sky, then simply pay their programming charges and bask in the glow of relatively cheap TV. The satellite development went rather well, and initial consumer purchases exceeded expectations. The systems had some initial bad press because they were unable to deliver local programming, and this was blamed on excessive regulation. A technical problem that exists with the system is something called rain fade, where a weather system standing between the antenna and its satellite can cause a weak signal, but in my talks with users of the systems, such problems are rare. This weakness is exploited by the cable companies in their advertising, which to me is a sign that the cable providers are scared.

Along the same lines, two companies named XM and Sirius envisioned a pay service for car audio. Once again by placing satellites in orbit, they planned on beaming compact-disc quality sound on a multitude of channels to subscription holders. XM radio went with conventional geostationary satellites to provide their signal, while Sirius opted to use an ellipsoidal orbit that took three satellites on a racetrack pattern around the Earth, staying at their highest point directly above the central United States. Each company bragged that their architecture had advantages over the other through their satellites' positioning, although when viewed through the entire country, it's likely that the systems are about even. Each system will respond quite differently to a single satellite failure, however.

The granddaddy of all the proposed satellite programs, however, was Teledesic. A joint venture involving Bill Gates of Microsoft fame, Teledesic's goal was to create a network of high-datarate satellites providing internet accessibility everywhere. Several constellations were discussed, but an early architecture called for 288 low-Earth orbiting satellites, all of which were pretty large. Through redesigns, the system got smaller and likely a bit more manageable, but at last word, the slowing rise in internet growth has pushed Teledesic back indefinitely.

A byproduct of these commercial satellite ventures was an interest in novel ways of replacing satellites in orbit. Several companies sprang up, partially to answer the X-Prize call (more on that later), but also to take advantage of the business opportunity provided by these satellite constellations. If one of the Iridium or Teledesic satellites were to fail on orbit, the idea was that these new companies could replace that single satellite quickly, instead of relying upon a rocket launch. Companies in this list included Kelley Space Lines, Kistler Aerospace, Pioneer Rocketplane, and Rotary Rocket. All of these companies proposed re-using their craft, and most of them were piloted. Each had its own unique approach for getting to orbit, including craft towed behind 747s until an engine fired to take them to orbital altitude, fly-back first stages that look like normal rockets, an air-breathing craft that takes on liquid oxygen from another aircraft before revving up a rocket engine to head to orbit, and a rocket with a number of spinning parts designed to simplify powered flight and landing. As the list of new satellite programs dwindled, so did the market for these delivery vehicles. Most of them are either scrambling to find new funding sources or no longer in business.

The X Prize was founded in 1996 with the stated goal of jump-starting space tourism, seen as one route to lowering launch costs. Founded by a space entrepreneur named Peter Diamandis, and based in St. Louis, Missouri, the prize is trying to restart the days of early aviation when rich men offered purses to inspire people to try for aviation records. The most famous of these, even though most people are more familiar with the pilot than with the prize, was the Ortig Prize that Charles Lindbergh claimed in his solo flight across the Atlantic. For a team to win the X Prize, they must build a vehicle that will carry three people to an altitude of 100km and return them safely. After this first flight, the same vehicle must repeat the mission 2 weeks later. More than twenty teams expressed interest in the prize, ranging from respected aviation pioneers like Burt Rutan to single people, supposedly working out of their basements or garages, looking for the fame that a first flight would bring.

The United States civil space agency, NASA, and the military, mostly represented by the Air Force, each faced a problem in that they wanted to decrease the cost of lofting payloads into orbit. The space shuttle hadn't met its promised price decreases, and military space launch remained more concerned with guaranteed access than cost in the early nineties, still recovering from the *Challenger* disaster. Each government agency took a different approach to the problem, however. NASA focused on reusable launch technology, while the Air Force tried to take an incremental approach to cutting the cost of space launch.

NASA's effort, inspired by the DC-X mentioned earlier, involved letting a series of contracts to refine designs on a vehicle demonstrator that would prove an operational concept. The concept vehicle was to be smaller than the envisioned final vehicle, and would fly several times through different flight regimes, starting low and slow and building to high and fast missions. None of these demonstrators were meant to achieve orbit. Once that concept was proven, the idea was that the contractor, seeing how useful the full-size vehicle could be, would choose to develop that full-size vehicle on its own, on a low- or no-cost basis for the government. Lockheed Martin was chosen to build the concept demonstrator, the X-33, and initially the company seemed bullish on demonstrating the concept and then pressing on spending their own money to develop the Venturestar, the X-33's fully operational big brother.

The X-33 ran into some technical developmental problems, and these were compounded by a cultural problem within the company developing the vehicle. Because the X-33 was supposed to be designed using both government and private sector money, most of the effort was done at very low cost. In many cases, hardware that was to be used for initial construction checks would also be used on board the vehicle, with no backup parts planned. This "protoflight" concept appeared to save a lot of money on the balance sheets before the hardware got built, but once hardware got built and tested, the savings vanished quickly. One such protoflight component was the liquid hydrogen tank. Designed to be ultra-lightweight, the tank was wound with carbon fiber, famous in tennis rackets and fishing rods, to give it strength. Carbon-wound metallic tanks work great for pressurized gasses where temperatures are not extreme, but liquid hydrogen is the second coldest substance known to humans, and the tank would be difficult at best to construct. After it was built, the unit went through a series of pressure tests to make sure that it could withstand the pressures that it was expected to see in flight. Normally, when more than one tank is produced, it's a good idea to test a tank to failure, which is a technical term for causing it to pop. When you've only built one tank that option doesn't exist, but I digress. During pressure testing, the team accidentally tested their only tank to failure. There was some finger-pointing as to whose fault it was, but things boiled down to the fact that NASA had contributed all that it was going to contribute to the project, and it was time for Lockheed to pay to get the program back on track. Nothing happened. Eventually, work stopped on the X-33, and it and the Venturestar program were quietly cancelled.

It's been said that Lockheed Martin was not greatly interested in completing the X-33 and Venturestar programs, and the argument makes some fiscal sense. Lock Mart, as it's known as in the business, was and remains one of the major suppliers of expendable launch vehicles in the United States. Expendable launch vehicles keep factories running to constantly build new rockets. If the Venturestar went into active service with the types of flight rates advertised, it's likely that only four or five could serve the needs of the entire world. Once those five were built, the manufacturing for the Venturestar and the expendable launch vehicles it replaced would be unnecessary. So, by making Lock Mart responsible for putting the Venturestar into service, the contract indirectly required Lock Mart to shut down their production facilities. This sort of activity doesn't sit well with investors, workers, or congresspeople.

The Air Force ended up with a better measure of success in its efforts to reduce launch costs, if only because hardware was actually built and flown. The service may not have reduced the real cost of launch that much. The Evolved Expendable Launch Vehicle or EELV was meant to take advantage of the lessons learned through decades of rocket launches and build new vehicles from scratch that were simpler and it was supposed cheaper to operate. When the contracts were let, the trend had more and more commercial launches flying on boosters, and the Air Force believed that the flight rate would be so high that the service's only "traditional" input to the project would be some start-up funds for initial development. After that, any US Government payload would simply have its ride paid for on the EELV, and the contractors in question would make enough money through their other customers that the rocket line would show a profit.

EELVs are designed to put a broad range of payloads into Earth orbit. If a small satellite needs to launch, the smallest version of the EELV is used. As payload size grows, additional rockets and capabilities are added, meeting the needs of the customer. By using the same components whenever possible, this design is kept relatively simple.

EELV, even though it's now flying, ran into two problems with its initial assumptions. First, it was expected that only one contractor's booster would be selected for operations. This stood to reason because, while space launches were more common, they likely weren't common enough to support two launch service providers. A series of spectacular and expensive launch failures in 1998-1999 soured the military's theory that one launch provider was enough. Since a launch failure grounds the booster involved for months, the military decided that it would be better to have boosters built by two companies "in the barn" so that one failure did not ground all operations. In a change of plans, two boosters were selected for full development. This cut the maximum market for each by 50%, but the military was willing to trade that for the ability to fly one booster if a failure grounded the other. The results of this decision were the Delta IV produced by Boeing and the Atlas V produced by Lockheed Martin.

The other assumption that fell apart in EELV procurement was the number of commercial launches that would be available to keep production rates high and prices low. The demise of big-ticket programs like Teledesic made for fewer launch requests by commercial vendors. This number dropped so low that Lockheed Martin even requested a release from its requirement to support a West Coast launch facility, which the government granted. In the last year, this decision was reversed, and Lockheed Martin will once again fly from the West Coast. To me, the initial decision to stop flights from California for one company is one of the few times that economics factored into a rocket decision, and it did so at the cost of the redundancy that the US Government fought for in funding two EELVs. Fewer launches meant that the real cost of each launch went up, so the contractors went to the government, hat in hand, saying that their EELV operations were not profitable at the current flight rate. The government took a step back to their regular ways of doing business here and wrote large checks to each launch provider to keep the production lines going. So, EELV flights will cost more than expected, but the actual cost per launch is not releasable outside of a planning number, such as $150M for the largest version of each booster.

Discussion of this timeframe in space travel would be incomplete without mentioning the largest envisioned project for the near future, the International Space Station.

Construction of the space station started in 1998 with the launch of the Russian-built, American-financed *Zarya* ("Sunrise") module. Next came the US launched *Unity* node, which will connect other modules to the station, followed by the Russian service module. Once that module was launched, the station was cleared for human occupation, and the first three-person crew took up residence. This moment was billed as the start of permanent human presence in space. With people there, each shuttle flight fit into an intricate web of missions taking place between Russian taxi cargo flights. While the cargo flights from Russia could provide fresh vegetables and clean socks, the space shuttle was required for the heavy lifting. Big missions included new components for the station, expanding its power sources and living spaces as time went on. In most cases, the missions proceeded without undue difficulty, although there were times where problems with shuttle hardware delayed launches and therefore delayed crew rotation. Through it all one thing could not be denied: the space station that NASA dreamed of for so long was coming into being, visible in the dusk and dawn skies as the brightest object in orbit. The station was coming together, even though it didn't live up to its original promises.

In 2001, the Clinton Presidency ended and a new one began, lead by President George Bush. Space was not a "hot topic" during the campaign, ranking somewhere below 20 as far as issues people were concerned about. During the 2000 campaign, I attended a debate between the science advisors for (then) Vice President Gore and (then) Governor Bush. I wanted to ask a question about space travel, particularly about setting a goal of a mission to Mars, but knew that such a question wouldn't be popular. I took a different tack, and posed the question as one related to education, asking if "pull" type of education incentives (exciting space efforts like a mission to Mars) were a better way to get kids interested in school than the "push" type of spending more money on education. The question got asked, but neither answer was that memorable.

A major change that the Bush Administration brought to light was that each government agency needs to prove its efficacy in order to maintain its funding. To meet this challenge, a high-ranking member of the Office of Management and Budget (OMB), Sean O'Keefe, proposed that NASA be held to a particular date (February 19, 2004) and a frozen budget to reach a certain milestone on the ISS. This milestone, called "US Core Complete" marked a time where, if desired, the US could then cut its support of the project and achieve peace with honor, having a functional space station on orbit. A couple weeks later, Sean O'Keefe was named as the head of NASA, and his plan was approved as the measurement used to gauge the space agency's success.

Meanwhile, construction of the International Space Station proceeded. Russian-launched, uncrewed cargo flights took supplies to the station as necessary, while Russian-built *Soyuz* vehicles cycled back and forth to keep an escape module at the station. On two occasions (and there are promises that there'll be more), these *Soyuz* vehicles carried paying customers, much to the consternation of the American space agency. The two tourists—Dennis Tito and Mark Shuttleworth—paid a lot of money to fly into space, and a lot of people fault them and/or the Russians for doing so. I don't. While I don't think that their flights were well timed as far as space station construction, I don't think that, left to their own devices, NASA would ever declare a "good" time for space tourists. Therefore, I applaud both men in their flights to orbit. In fact, I got a chance to thank Dennis Tito in person, when he spoke at The National Air and Space Museum in Washington. Now let me be clear: I don't think that space tourist jaunts to government-purchased orbiting platforms are the right way to start a space tourism business, but I am glad that someone opened the door to space tourism, adding the term to our lexicon so that discussions of other ways to do it became more serious than before.

REALITY STRIKES AGAIN, NEW HORIZONS: 2003

On February 1st, 2003, my son and I were visiting my parents. My wife called and told me to turn on the TV, because it looked like the space shuttle *Columbia* broke up on its return to Earth. Partially because I had my son with me, and I didn't want to have to explain a particularly grizzly scene that appeared on the screen, and partially because my opinion of the media has changed significantly since the *Challenger* accident, I didn't spend the entire day glued to the television. I find explaining things to a three-year-old to be a refreshing exercise in boiling things down to the important facts. I told him, "I'm sad because seven people went up in a space ship, and they won't be coming home." There's a whole chapter coming devoted to this day and its fall out; as horrific as the event was, it was not all that happened in 2003.

Space: What Now?

The X Prize, a $10M purse designed to open the realm of space tourism through commercial means, discussed earlier achieved its fundraising goals early in the year. The funding method, though unconventional, will pay a winner as long as they fly before the end of 2004. In the agreement, an insurance company will pay the $10M to a team that achieves the stated goals of the contest before the end of '04. If no one achieves the contest goals, then the X-Prize foundation will pay the company the funds raised so far, said to be about $3M.

Soon after the fully funded announcement, Burt Rutan, the aviation pioneer who built the *Voyager* aircraft to fly around the world on one tank of gas, unveiled his competition entry. A carrier craft dubbed "White Knight" carries a smaller capsule called "Spaceshipone" to a normal cruising altitude for an airliner. At the appropriate time, *White Knight* drops its charge, which then fires a rocket engine (burning a mixture of rubber and laughing gas) to then coast to the required altitude of 100 km. After its engine fires, *Spaceshipone*'s wings fold up to keep the craft stable throughout its flight that carries it up to the very fringes of the atmosphere and provides the passengers several minutes of weightlessness. As it falls back to Earth, the folded wings hold the capsule steady in such a way to minimize heating on the craft. When it returns to an aircraft flight altitude, the wings are folded back into their normal position, and *Spaceshipone* glides to its runway. Throughout late 2003, the system flew several times, experiencing problems on some flights, but fixing them and moving on. On December 17th, the pilot ignited the rocket at the back of *Spaceshipone* and broke the sound barrier while climbing to an altitude of 21,000 meters, or 68,000 feet. On this flight, the left landing gear gave way as the craft glided to touchdown, but the team says the damage is easily fixable. On this same date, Paul Allen, cofounder of Microsoft, confirmed rumors that he was the financial backer of the effort.

While Rutan is the current media favorite of the X-Prize contenders, he is not alone. Another outfit, called Armadillo Aerospace, is based in Texas and also has some flying hardware. Their design is based on hydrogen peroxide rockets, and is backed by a software millionaire, John Carmack. They also have hardware built, though their website focuses on small-scale piloted work such as hover vehicles and the like.

Along the lines of internet millionaires who still hold that title, there are several efforts underway right now to get people or materials into space, funded by the recently very rich. During this timeframe, other companies such as SpaceX and Blue Origin, mentioned earlier, made their plans wildly public and partially public, respectively.

In the summer months of 2003, three rockets launched from the planet Earth sent craft on their way to the same objective—Mars. Another spacecraft, launched in 1998 from Japan, flew past Earth to make its final course change designed to take it to Mars around the same time as all the other craft. Combining these four craft (some say that there are actually five, since the European *Mars Express* craft is both and orbiter and a lander) with the two already in orbit around Mars (the *Mars Global Surveyor* started orbiting the Red Planet in 1997, and 2001 *Mars Odyssey* arrived in the year of its name) there will be up to 7 (again, depending on your accounting methods) spacecraft from Earth exploring Mars in January 2004. After arrival, the numbers were cut down a bit. The European lander, *Beagle* 2, was never heard from after its expected landing on Mars. The Japanese spacecraft, long delayed in its arrival at Mars, eventually succumbed to a series of electrical failures and did not enter Mars orbit. Let's take a look at the craft and their capabilities.

The United States has arguably the best chances of returning a wide variety of scientific data, due to two factors. First, they are the nation with the most experience in Mars exploration (even though this didn't help them much during the failed 1999 missions), and second they have returned to previous practices and decided to send identical spacecraft on the same mission. This was common practice in the past. In some cases, it meant that two spacecraft would visit the same planet at different times such as *Pioneer* 10 and 11 and *Voyager* 1 and 2. For other missions, it meant that a big failure on the part of one mission meant that the program still got some data. Several of the *Mariner* spacecraft were launched in pairs, to have only one of them reach the goal. In 2003, two identical Mars Exploration Rovers were launched to Mars. Through a nationwide naming contest, the names "Spirit" and "Opportunity" were chosen for the rovers, and they arrived successfully in early and late January.

The rovers are an amazing feat of engineering. About the size of a washing machine, they're designed to travel up to 1 kilometer on the Martian surface, yet can fold into a pyramid shape for their 200-day trip from Earth to Mars. Upon arrival, they bounced along the surface for up to ten minutes, then unfolded and went about their business of exploration. They carry out that exploration mostly through a set of instruments mounted on the end of an arm at the front of the rover. The instruments include a rock polisher, a magnified camera, similar to a field glass that a geologist would carry, and two types of spectrometers, designed to rest against the rock and determine what the minerals are inside it. One spectrometer, called the Alpha Proton X-ray Spectrometer or APXS, is designed to give a general impression of the makeup of a rock at the elemental level. Basically, it tells how much of a particular atom is found in a rock. The other instrument, called the Mössbauer spectrometer, describes the types of iron minerals found in a sample. While this is more focused than the APXS, the rovers are being sent to areas where iron is thought to be found that was affected by water over time. The rovers have lasted much longer than expected, showing some signs of wear only after weathering the worst part of a winter on Mars. Their finding of hematite as well as mineral salts have reinforced the idea that parts of Mars were covered by an ocean in the past. The rovers are not designed to search for either current or past life, but in the unlikely event that there is something relatively large within the rocks, an accidental discovery is not out of the question. The prime focus for these rovers is to look for minerals that have been affected by water throughout Mars' history.

The European spacecraft orbiting the Red Planet right now is named *Mars Express*. Designed and built on a relatively low budget for such science missions, *Mars Express* carries a suite of instruments that will orbit Mars taking pictures and using active radar to explore the surface. Throughout 2004, both forms of information were coming in, with radar data confirming earlier US claims that water exists on Mars just below the surface, and spectacular pictures being returned as well. The big gamble for *Mars Express*, and its European builders, came on the surface, though.

As *Mars Express* approached Mars it released a small capsule, designed to land on the surface of Mars. The payload of this capsule, called *Beagle* 2, is the size of a medium suitcase, and it was meant to unfold once it bounced to a stop on the surface, exposing solar panels to the sun. This craft is not mobile like the US rovers, but it has a group of instruments with some amazing possibilities. Like the US vehicles, the instruments on board *Beagle* 2 are at the end of a movable arm, so they will be able to sample rocks that are nearby. Included instruments are magnifiers, spectrometers, and rock polishers. The most innovative instrument onboard the *Beagle*, however, is a device called "The Mole," which can actually crawl across the surface and burrow down into the dirt. Once under ground, it can capture a sample of material where it is and bring it back to the craft for further analysis.

The Japanese craft, named *Nozomi*, or "Hope," was launched during a much earlier launch opportunity (see Chapter Eleven, the technical stuff). Unfortunately, the spacecraft suffered a pretty major malfunction of its propulsion system and was unable to send itself on its planned route to Mars. As a tribute to the ruggedness of its design and the ingenuity of its team, this loss of propulsion didn't mean the end of the mission, because they still had the propulsion ability to bring the craft back to Earth, swinging past to

57

increase its speed. This changed timeline would get them to Mars, although it would be years after their original plans. In another unfortunate incident, a solar storm took out some portion of their communication capabilities. This forced some changes in how they were going to beam data back to Earth, but didn't affect their ability to arrive at Mars. Had the spacecraft arrived successfully, it would have explored Mars' atmosphere interaction with the sun.

In early summer 2003, a bright dot appeared in the night sky. It didn't just show up there, of course, but it was definitely something different than the stars nearby, and it was getting brighter all the time. In July, this red dot made a close approach to the moon, actually going behind our nearest neighbor as viewed from some places in the United States. This created a spectacular opportunity for astronomers around the world, but few people with interests outside of space noticed. As August drew near, however, a bigger event was reported in the popular press, which specializes in superlatives. As it turns out, on August 27th, Earth and Mars would be closer together than they'd been in 60,000 years. How much closer, you ask? It turns out that Earth and Mars are, on average, about 225 million kilometers apart. On August 27th of 2003, they were 54 million kilometers apart. This means that anyone could see the polar caps of Mars using a commercially available amateur telescope. My father received an e-mail with the somewhat deceptive title "Mars to appear as large as the moon in August." As I read the posting, it turns out that they were saying that if you were looking through a modestly powered telescope, then Mars would appear as large as the full moon. OK, I'll grant them that. The event did catch people's attention, though, and most people I talked to in the National Air and Space Museum for the following weeks wanted to know more about what was going on.

So, what was going on? Well, Earth and Mars both orbit the sun. As the recent comedy series put it, Earth is the third rock from the sun, and using that logic, Mars is the fourth. While Earth orbits the sun in 365.25 days, Mars takes 687 Earth days to complete one transit. This rate difference means that Earth and Mars come "close" to each other about every 26 months. In one of these close approaches, Earth passes Mars "on the inside" and the term for the moment when the sun and the two planets form a line is opposition. It's usually about three months before this close approach that Earth launches its space probes to Mars, but there's more on that in the technical section. Now you're probably asking, if Earth and Mars are close together every 26 months, why was August 2003 such a big deal? The reason is that while Earth's orbit is very close to a circle, Mars' orbit is much less so. If you looked at the solar system from above, Mars' orbit would look slightly egg-shaped (the proper term is elliptical, which I have to include for those who are reading this book looking for technical errors, but we'll move on with egg-shaped). This shape of an orbit means that Mars is relatively close to the sun for part of its orbit, but then, ½ of a Mars year later, it is relatively far from the sun. You'll see that the 26-month period between Earth and Mars close approaches is not the same length as Earth's or Mars' year, so that means each time one of these close approaches takes place, Mars is at a different distance from the sun. In extreme cases, such as the 1995 opposition, Mars was very near its farthest point from the sun, and the Earth-Mars close approach was an unremarkable 105 million km. Eight Earth years later, when the Sun and Mars opposed each other as viewed from Earth, Mars was only 54 million km away. This translates to an incredible show here on Earth, with a shining beacon in the night. When I look at it, I believe I understand why people spent so much time trying to relate celestial events, such as a Mars opposition, to events here on Earth. After all, look how much excitement this event brought. Even if you never looked through a telescope before, you probably heard something about the opposition, or at least looked up at night and wondered what the bright star was.

The military launch business saw good times and bad times in 2003. Operationally, things are going well, with both EELV boosters flying their first and subsequent missions successfully. The next big thing up for these rockets is a demonstration of the Boeing heavy version, a requirement before the government will put one of their high-priced satellites on top of the craft. The bad times came in the form of litigation between Boeing and Lockheed Martin in their development of the EELV. It became public in 2003 that, during design, building, and most importantly pricing of the EELVs, Boeing had an employee who formerly worked for Lockheed Martin. Apparently, this person used the fact that he worked for Lockheed as a point on his resume, and either explicitly or implicitly mentioned that he also had lots of documentation from Boeing's competitor as to how they planned to price their boosters. This information translated to a huge advantage for Boeing, and, according to government findings[39], they used it to win the lion's share of launch business from the government. In response, the US Air Force (representing the US Government) took several missions that were formerly supposed to fly on Boeing boosters and moved them over to the Lockheed Martin column. In the most recent awarding of new contracts, Lockheed Martin carried the day with the most launches. Perhaps the biggest revelation, however, is that the government changed the terms of Lockheed Martin's contract to allow them to build a launch facility on the West Coast. As mentioned previously, the lack of commercial interest in launching from

Vandenberg and the few Air Force launches from that coast created a situation where Lock Mart found that it didn't make good business sense to open a launch facility there. Boeing now states that they've learned their lesson and they'll never do it again.

The US war to depose Saddam Hussein in March of 2003 again pointed to the critical role that satellites play in modern military operations. In the previous Gulf War, the Global Positioning System (GPS) was incomplete, giving only occasional positional coverage to the region. During operation Iraqi Freedom, GPS satellites surrounded the globe and provided constant three-dimensional coverage to the field. Other systems, used to warn military members and civilians alike of the dangers of missile launches, were not fundamentally improved but were used much more efficiently, cutting response time and increasing the chances for interception of the incoming missile. Some space infrastructure limitations were brought to light, however, as the newly popular uncrewed aerial vehicles (UAVs) ran into difficulties flying not because of maintenance problems, but because there wasn't enough data transmission capability through the theater. UAVs beam back a lot of information, and that information needs a pretty big pipe to get through properly.

Chapter Three

The *Columbia* Tragedy

This chapter is based on the work of the Columbia Accident Investigation Board. I was waiting online at 10 a.m. on August 26th 2003 to download volume 1 of their report, and I grabbed it as soon as I could because I expected the website to overload quickly. I got my copy, and never heard about any problems with the website. Oh well.

Before covering the destruction of *Columbia*, a moment discussing the mission itself is in order. *Columbia* was flying a dedicated science mission, scheduled for 16 days on orbit to run a battery of tests on crew members, as well as covering a number of scientific investigations into observations from space, combustion, and animal behavior in space. This flight was designed to fill the gap between earlier "science mission" trips into orbit and when the space station would be fully operational and be able to take its place as the primary location for scientific inquiry in low Earth orbit. While there's been some discussion of the type of experiments chosen to fly on this mission and their overall importance, *Columbia*'s flight was meant to be another, fairly routine trip for the shuttle. Of course, it was anything but.

The Columbia Accident Investigation Board report was published between August 26th (volume 1) and October 28th (Volume 2-6). This chapter is based almost exclusively on Volume 1. Volume 1 contains just under 250 pages, and is divided into three parts, discussing the accident, why the accident occurred, and the way ahead. If you find the information in this chapter interesting, I recommend that you read the entire CAIB report, Volume 1, for more details.

First, let's introduce the space shuttle system. The space shuttle, also known as the Space Transportation System, is made up of three parts at liftoff: the orbiter (the winged vehicle that returns to Earth, embodying what most people think of when they think of the space shuttle), the external tank (the large brown tube which the orbiter sits astride at launch—it is the only part of the vehicle that is discarded on each flight), and the solid rocket boosters (there are two, one on either side of the external tank). When they are all together at rollout to the launch pad, or during launch itself, these components are sometimes referred to as the shuttle stack. Throughout this chapter and the book, I will use the term orbiter and shuttle interchangeably, even though I'm sure some space purists will not like the practice.

JANUARY 16 – FEBRUARY 1, 2003, 08:44:09 EST

On the morning of January 16th, 2003, seven US astronauts—Michael Anderson, David Brown, Kalpana Chawla, Laurel Clark, Rick Husband, William McCool, and Ilan Ramon—were awakened for their mission on board the Space Shuttle *Columbia*. At 10:39 that morning, East Coast time, their mission began atop a pillar of smoke and flame to orbit. Initial indications were that the *Columbia* took her crew safely to orbit, though later data review would show this not to be the case. It turns out that at 81.7 seconds after launch, a piece of foam, about the size of a cinder block, separated from the brown external tank. The foam came from an area known as the bipod, where the tank connects to the front of the orbiter. Because the foam was so light and the air it broke off into was moving so fast, the wayward hunk was blown backward into *Columbia*'s wing at a speed somewhere between 190 and 260 meters/sec (416 to 573 miles per hour). At impact, the foam likely cracked or literally punched through the protective coating that covers the front of *Columbia*'s wing, but no one knew this for some time. Later, as part of the investigation, the accident board found that a sensor recorded a small temperature rise near the suspected impact site as the orbiter climbed into orbit. This data wasn't available to the ground, but was recorded on board *Columbia*.

About an hour later, *Columbia* made its final maneuvers to settle into earth orbit—and then the crew set up house for their 16-day mission. The first day was spent largely unpacking and setting up future equipment, but then the team split into two teams (blue and red) that would allow around-the-clock operations and maximize useful science from the mission.

Over the next two weeks, the crew carried out a range of experiments, including:

1. Measuring dust storms over the Mediterranean Sea and Sahara Desert. This same instrument captured the first images of rare weather events called sprites
2. Studying the fluid properties of Xenon
3. Experimented with different methods of fire suppression in microgravity
4. Measured the body's response to microgravity, especially in the areas of protein manufacturing, renal stone formation, and saliva changes due to viruses
5. Took images of meteoroids entering Earth's atmosphere.
6. Experiments suggested by school children were carried out and the results were beamed back to Earth directly to the students in the project.
7. Produced bone and prostate cancer tumors the size of golf balls, the largest such samples ever created in microgravity.
8. Generated the lowest-power flame ever, producing 1 watt of energy. A birthday candle, by comparison, produces approximately 50 watts.

A few special events happened during the flight. One was related to the launch event and future occurrences, some were simply convenient news opportunities, while others were ironic in nature. First, on day two, a piece of debris drifted away from the shuttle. No one knew this at the time, and its existence didn't even come to light until after the orbiter broke up on entry and the Air Force reviewed its radar data. More on this later. On at least one occasion, *Columbia* passed within sight of the International Space Station, then crewed by three people and appearing as a bright, very fast-moving star. *Columbia* and the station were in very different orbits (more on that in the technical section) so were in no danger of striking each other, but it did provide for the largely symbolic exercise allowing both crews to talk directly to each other over their radios. Then, on January 28th, the crew took some time, and along with mission control, held a moment of silence for the crews of the *Challenger* and *Apollo 1*, two other space missions whose crews were lost.

The mission was not perfect, and several glitches prevented the crew from doing everything on their agenda. Early on, one of their centrifuges, a device used to spin biological samples and separate them into various layers, stopped working and forced the crew to shut it down. There were also climate control problems in the science lab, causing water to pool under the floor and in other inconvenient places. This led to changes in temperatures on board the lab, trying to control the amount of water that condenses out of the air. These problems were annoyances more than anything else, and the majority of the mission's objectives were met.

On the last day of their mission, the crew packed up supplies and got ready to return to Earth. They got into their bright orange suits and settled in for what was expected to be an uneventful ride back from orbit to cap off a largely successful mission. At 8:15 EST, they fired small rockets that slowed their orbital speed by a mere fraction of its value, and then sat tight to watch the Earth grow larger in their windows and welcome them home.

08:44:09 EST – 09:05

At 08:44:09 a.m. eastern time, the *Columbia* passed 123,000 m (400,000 feet or about 75 miles), which is a mostly arbitrary point known as entry interface or EI. This point describes the time when the atmosphere begins to affect the spacecraft, so the first tenuous wisps of air were moving over the craft and finding their way through the damaged left wing leading edge into the wing itself.

Even though there was airflow, the first recorded event that was "off-nominal" or "different than expected" took place at EI + 270 seconds, where a sensor indicated a slight deformation in the metal it was mounted to. Ironically, this event was not broadcast to the ground, but was recorded on board *Columbia* in a device known as the Modular Auxiliary Data System, or MADS. Think of MADS as a flight data recorder used on board commercial aircraft to record the last moments of flight of the aircraft. The difference here is that the MADS was intended to provide data for post-flight analysis after a nominal flight. It was not designed to survive a crash, but it did so quite well.

The second off-nominal reading came at EI + 290, where a temperature sensor (the same one that recorded the small temperature rise on ascent) located very near the suspected foam impact site showed a slow increase in temperature that wasn't normal to a reentry. This sensor did not show very much temperature change, and it's suspected that the area near the sensor was not exposed to direct heating, but something called "sneak flow."

The earliest concrete evidence of a burn through, or hot gas working its way through the wing, started at EI+425 seconds. Here, a temperature sensor showed a small temperature rise that quickly ramped up to higher and higher levels before losing data at EI+525 seconds. This sensor actually read a temperature of 450 degrees, the maximum it was designed to report, just before all data was lost from it.

While all this was going on, it's ironic that parts of the shuttle were actually heating less than on a normal return from orbit. The left side of the orbiter and the bottom edge of the orbital maneuvering system (OMS) pod were warming at lower rates than previously experienced, likely because of the change in airflow around the left wing.

At EI+500 seconds (8 minutes and 20 seconds), the fact that the left wing was interacting with the air in a drastically different way than the right wing began to impact how the shuttle was flying. Essentially, the orbiter was trying to turn left more than it should. The shuttle is on auto pilot during this time, with the computers running the show, and the small jet firings that were taking place to keep the craft stable would not be remarkable on board or on the ground.

The ground got its first indication of a problem at EI+530 seconds. Here, the information wasn't a sensor saying that a temperature was high or metal was deforming, it was the loss of data from a sensor. What happened is that the wire delivering data from that sensor was cut by either the ultra-hot flame moving through the wing or by deformation pulling the wire a apart. Loss of sensor data is unusual, but it's always associated with a maintenance chore after the flight and isn't something that leads to immediate alarm. At EI+562 seconds, another bundle of sensors failed.

By this time, the orbiter was over the continental United States, and people on the ground, out watching the reentry, could see (and videotape) debris coming off the vehicle. They reported and recorded seeing a number of pieces coming off *Columbia* throughout the rest of its journey. In some cases, the debris events were linked directly to events recorded on board.

The hot gasses in the wing reached the left main landing gear wheel well by EI+601 seconds (about 10 minutes) and this relatively open area started to heat up quickly. This initial rise is attributed to the well being warmed from air outside it, not by the superheated air actually getting into it, which happened later.

To give an idea of how rapidly things were changing, at EI+602 seconds, the orbiter stopped trying to turn to the left and started trying to turn to the right. The theory for this is that the left wing changed shape enough that now it was producing more lift than it should. Again, the computer in control at the time simply changed the jets that it was firing, and could still maintain control, so nothing really noticeable changed for the crew or the ground.

More components in the wheel well showed increasing temperatures between EI+727 and 790. Hydraulic lines and the huge, heavy tires started getting warmer for a short period of time before telemetry was lost. The fact that the tires, which are known to be able to absorb a lot of heat before showing a significant change in temperature, could warm up at all in so short a time indicates how hot the gasses warming them were.

During this time, the *Columbia* was scheduled to turn as part of its normal entry plan, and it did so. These turns allow the craft to bleed of extra speed, and also let it direct its flight path for a precise landing. I find it amazing that the computer could still maneuver the craft at this point, given the severely damaged left wing, but the turn went well, with only slight differences from what would have happened without the damage.

The left wing apparently changed shape rather radically at EI+834, because the computer made another adjustment to the flight control surfaces and number of jet firings. Film footage from the ground also confirmed that several pieces of *Columbia* broke off at this time.

At EI + 897, the left landing gear gave an indication that it was down and locked, prepared for landing. This was a single source of data, and other data that would corroborate the fact did not appear. The other data that tells us that the gear was not down includes information saying that the landing gear door was still closed, and another sensor showed that the landing gear was still up. The board concluded that the down-and-locked telemetry came about because a wire was burned through.

The next series of events happened quite rapidly. Another major shift in the vehicle's shape took place at EI+917, and mission control at Houston lost signal with the orbiter at EI+923 seconds. The orbiter lost its ability to hold attitude control at EI+938. By this time, the left wing was so damaged that no matter which jets fired, *Columbia* could not point in the direction it wanted to. The orbiter was traveling at 17 times the speed of sound at this point, and traveling at that speed requires some very precise pointing of the craft. Once *Columbia* drifted from that precision pointing, its structure could not hold up for long. The last piece of data from the MADS was recorded at EI+970 seconds. By this time, mission control had gone into a contingency search mode, looking for any trace of *Columbia*, and television news stations were gearing up to start complete coverage of the event.

THE SEARCH FOR ANSWERS

The investigation into the *Columbia* accident began soon after the last pieces fell to the ground. NASA administrator Sean O'Keefe was at Kennedy

Space Center to watch the orbiter's return at 9:16 a.m., but knew something was wrong when he didn't hear the telltale sonic booms that announce a shuttle's return. At 10:30 a.m., he activated the International Space Station and Space Shuttle Mishap Interagency Investigation Board. As it turns out, this group of people was already designated just in case of such an accident, and was made up of people who held specific roles spread across the US government. Retired Admiral Harold W. Gehman, Jr., was appointed as the board's chair. This team was notified of the event by noon, and members were instructed to meet the next day, February 2nd, at Barksdale Air Force Base.

Debris recovery started almost immediately, with NASA asking the public not to touch any pieces of the shuttle that they found. While some people questioned this guidance, saying that most of the shuttle parts were not hazardous, I believe it was the right thing to say. It's true that most of the shuttle parts are not hazardous, but how can a person on the ground know which are hazardous and which aren't? Is it wiser to produce a big list of possible debris, saying which ones are dangerous under which conditions, or just to say, "stay away from them all"?

The board got started by looking into the sources of data they had for their investigation. These sources included visual/tracking (launch footage, breakup videos taken from the ground, and other images and radar reports taken throughout the mission), on-board data (including data that was beamed to the ground as telemetry and that which was stored on board in the MADS), debris found on the ground (based on where it was located and the condition it was in), and tests run on the ground using shuttle hardware, computer modeling, and any other means necessary to gain insight into what happened. Very important data was also gathered from interviews with shuttle team members, gaining a fuller picture of the *Columbia* tragedy as it fit into the larger picture of what was going on with NASA.

The first, most critical, and time-sensitive data to gather was the debris leftover after the breakup, and search teams were created to sweep the states of Texas, Utah, Nevada, and even California to look for pieces of *Columbia* that might provide a clue. Each part, when it was found in the field, had its position noted as well as its condition. It was then gathered up and sent to Kennedy Space Center, where a ghostly reconstruction of the orbiter took place on the floor of a dedicated hangar. The crew's remains were found in the first hours of the search and returned to their families. The cause of death for all crewmembers was blunt trauma and hypoxia (lack of oxygen).

A call was made for people who'd given into their whims and gathered souvenirs from the tragedy. Amnesty was offered for those who came forward, and penalties threatened for anyone who did not surrender their finds. Over the months, approximately 38 percent of the orbiter, by mass, was recovered. Important finds within the debris field included pieces of the left wing, as well as the MADS recorder, used to reconstruct much of what happened.

This type of investigation is very difficult, and is almost always based on incomplete information. It's the job of the accident investigation board to find the data that they have, determine what they need, and try to fill in the gaps with analysis and experimentation that fits with all the information that they have. The task is often compared to doing a jigsaw puzzle with only some pieces and only a vague idea of what the final picture is supposed to look like. About the only thing this type of investigation can be compared to is an airplane crash investigation, but there are many problems with this comparison. In most cases, a large portion of the aircraft is recoverable (although in a radically different form than they were when the craft was flying, to be sure) and the forces involved in an airline crash are much better understood (through the unfortunate experience we have in dealing with them) than was the case with *Columbia*.

While the debris search continued, the team set about working with the data they had in hand, which took the form of visual images of the foam strike, radar information from the Air Force, video images of the reentry, and information beamed to the ground as telemetry. Again, these pieces of data showed radically different views of the event, and it was the board's job to find a sequence of events that fit the data they had, without supposing too much else.

The video of the foam strike was used to try and calculate the amount of damage that was possible in the captured event. Here, the imagery wasn't as good as was possible, because the best camera available to film on launch day followed the shuttle in its flight, but was out of focus. So, the board relied on other footage that the public had seen just days after the accident. There were two sources of this video data, one camera that was 17 miles away from the stack when the strike occurred, and another that was 26 miles away. Using these grainy frames, the board determined that the piece of foam that struck the forward edge of *Columbia*'s wing was moving between 200 and 250 meters per second, or about 400 to 600 miles per hour, and was between 21-27 inches long by 12-18 inches wide. They also determined that the piece of foam struck within a fairly small area on the left wing. The data gathered from the video footage matched a computer model created as part of the investigation. Some wonder how a piece of foam could strike the orbiter moving so fast, when it broke off the external tank mere fractions of a second before. It turns out that the speed is due to the foam and its properties. Before giving an example, I guess I need to say that you should not actually carry out these experiments.

Think of riding down the road at highway speed, and throwing a rock out the window. The rock will essentially travel along with the car, arcing away from the vehicle until it strikes the road, and then it will start to slow down significantly. Now, imagine throwing a styrofoam cup out the same window. The cup, being much lighter, but having a much larger surface area, will appear to you to quickly race backwards, even though it's actually slowing down. The rock and cup have radically different ballistic coefficients, or ratios of how much they weigh vs. how much surface area they have. These effects are visible at mere highway speeds, but the shuttle stack was moving much faster when the piece of foam came off the external tank.

While determining the speed of the foam, the board also worked to show where the foam struck *Columbia*. The leading candidate strike zone was near the leading edge of the orbiter's left wing, near where the wing changes angle. The board was pretty sure that the foam struck on the tank side of the orbiter, because there was no visible debris that flew over the top side of the wing at impact. The exact location was critical to the investigation, because the material on the very leading edge of the wing, called reinforced carbon-carbon (RCC), is very different than the material that covers the majority of the orbiter's skin, the silica tiles that people who follow shuttle missions heard so much about in the early days. The RCC is a very hard but somewhat brittle material that's designed to hold a shape (in this case, the front of the wing) with very little structure behind it. In fact, the area directly behind the RCC is nearly empty, with each RCC panel held only at its top and bottom. The silica tiles, on the other hand, are also fragile but are mounted directly to the orbiter's skin. This gives them a measure of support that makes them a little more resistant to impact damage. Because of all the attention these silica tiles received during the early shuttle missions, NASA's experience with them is much greater than with the RCC.

On day 2 of *Columbia*'s mission, a piece of the shuttle actually "fell off" and was tracked by US military's space surveillance network. Some might ask why a military radar track of an object coming off the shuttle in orbit didn't raise an immediate alarm, and there's a reason. The orbiter in its natural habitat, space, is not a very clean vehicle. Some materials, like wastewater, are dumped overboard routinely. These materials tend to freeze right outside of where they're dumped, and then break off sometime later. In some cases, military radar can pick up these pieces. It wasn't until after *Columbia* broke up that military personnel reviewed the tracking data and found that the piece they tracked was not a typical piece of debris. It's likely that all debris coming from the shuttle will be analyzed with a bit more scrutiny in the future. The accident board, using the radar tracking data, determined that the piece of debris was most likely a piece of *Columbia*'s left wing, although there wasn't enough data to state exactly what the piece was.

Video images of the reentry were gathered, starting at the California Coast and continuing eastward along the *Columbia*'s flight path. They all showed disturbing "debris events," which is an accident investigation board term for "pieces coming off the spacecraft." The quality varied greatly from handheld video cameras to hobbyists at Kirtland Air Force Base taking images through a store-bought telescope, to actual trackers assigned the job of recording the reentry, and they were all put together to paint the entire picture of *Columbia*'s last moments.

Measurements taken on board *Columbia* and then beamed to the ground, known as telemetry, was also analyzed to find what indications mission control had for any problems on board. As it turns out, telemetry during reentry fades in and out, as the orbiter is surrounded by superheated air. These telemetry blackouts are short-lived in a normal mission, but that meant that some data from the craft would not be available on the ground. Also, as *Columbia* lost contact with the ground for the last time, some data was beamed down, but ground processing computers, seeing that the data had holes in it or wasn't assembled properly, didn't display that last information to the ground. During normal operations, ratty data can lead to bad decisions when the ground reacts to a problem that isn't real, so this filtering is a good idea. In the case of *Columbia*, that data could have critical information in it as to what was going on in the orbiter's last moments. The board mounted an effort to get the last bits of data, and try to sort through what was good and what was bad. This additional information helped a bit, but its significance was diminished when the MADS was found.

Analysis of telemetry received on the ground reaffirmed what most people already knew. The main indications of problems on the ground, up until the loss of communications with the orbiter, were losses of information from sensors in the left wing. Data dropouts like these are not common, but they typically do not indicate a serious problem on board, either. Also, by the time telemetry started dropping out on the ground, the fate of *Columbia* was sealed, the damage done to the left wing was unrecoverable.

Debris found in the field was coming to the board at a steady stream by mid-February, and there was an obvious difference in the amount of debris and the damage level in the debris between the left and right wing. There was much less of the left wing found than the right, and many of the parts that were found showed tremendous heat damage. Pieces of the right wing were found much closer together, with less heat damage. One extreme example of this difference was the landing gear tires. On the right side, the tires were found practically intact, while the left sides were severely melted. This information matched the theory that *Columbia* attempted to return to Earth with a damaged left wing, which matched with the video footage showing the foam strike on lift off.

Using the few pieces of left wing available, along with the MADS recorder, the team set out to verify that their theories were correct. This doesn't mean that they'd discounted other ideas of what happened out-of-hand, but the evidence was overwhelming at this point that they were on the right track. Fortunately for the investigation, orbiters have some very distinctive metal in the leading edge of their wings, right behind the RCC panels. If the area behind the RCC panel was exposed to extreme heating, these unique metals would melt and spray around to other areas of the *Columbia*. The team looked for this metal spray and found it in multiple locations. Once a molten spray was confirmed in the debris, a model of the orbiter was placed in a wind tunnel to see what specific kind of damage would cause the superheated air flowing past the wing to pick up some metal and spray it onto the orbiter.

By far, the most important find in the debris field was the MADS recorder. It noted information that wasn't included in telemetry beamed to the ground. It was hoped that this recorder continued to log information even after the signal was lost with the ground, though this would be dependent on the manner in which systems failed as the orbiter broke up, and wouldn't be known until the data was played back.

Information from the MADS recorder further confirmed damage to the left wing by showing temperature rises on metal parts within the wing itself. On one sensor, located right behind the (suspected) missing RCC panel, the temperature rise was almost immediate. Other sensors showed where metal was bending, warping, or otherwise changing shape as the wing lost much of its structure. In many cases, the data only varied from a normal reentry for a short time before the wire connecting the sensor to MADS was cut by the blowtorch-like air, and data was lost. Ironically, the MADS data showed a small abnormal temperature rise during launch on the wing directly behind the damaged RCC. Again, the usefulness of the data is questionable since no one had it on the ground in real time to make a decision based on it.

Following this trail of different-than-usual followed by nonexistent data, the board traced the most likely path of gases through the interior of the left wing back to the landing gear bay.

Other sensors monitored by MADS, located along the side of the orbiter, noted different temperature signatures than was recorded in other missions. These measurements confirmed the different airflow that was shown by deposits of metal found in recovered materials from proper sides of the orbiter.

MADS data provided a more complete picture of the final seconds of the orbiter, because it continued to record after *Columbia* stopped broadcasting to the ground. MADS data stopped at EI + 970 seconds.

Seconds later, *Columbia* was gone.

NASA's Response During the Mission/Missed Opportunities

One of the most frustrating parts of the *Columbia* breakup is to see how many things went wrong, both mechanically and in policy/practice wise, to cause the dark day in February 2003. Like the case of an airline crash, it's almost never one event that causes a disaster, but one event surrounded by a multitude of small assumptions and mistakes will definitely cause an accident. Seeing them spelled out, in order, as part of the accident investigation board report makes the situation look even worse.

In the report, the board uses the term "Missed Opportunity" to describe a moment where, if someone had done or said something different, more data would have been gathered that could have changed events. There were seven total missed opportunities.

First off, a critical point: no one knows of a sure-fire method or action that would have brought *Columbia* home. Any discussion here is purely Monday-morning coaching, second guessing others in the decisions they made at the time. It is much easier to list a string of events after the fact and say that someone doing something differently would have changed the course of history than it is to be a part of events as they unfold and know it's time to speak up and change course.

The foam strike now accepted to be the cause of the *Columbia* disaster was not widely known about until the second day of the mission. Personnel on the ground typically review launch images on the second day, and the strike was pretty obvious on the video footage that they had. Concern levels varied almost uniformly between managers and engineers. Managers felt that the debris strike was within the experience base, meaning that since the orbiter had survived similar strikes (their data for concluding that the strike is similar to previous strikes were not specified) this one was survivable as well. Engineers, however, wanted to gather more data immediately to ascertain how serious the impact was. An entry in the mission evaluation room manager's log shows an immediate bias towards low concern. Some people who saw the event became alarmed and started pressing with an effort to try and determine the severity of the impact through both analysis and through a request to get imagery of the *Columbia* on orbit. Both actions took place in parallel, with the first informal request about imagery actually taking place on day 2 of the mission.

Analysis of potential impact damage took place using a piece of computer software called Crater. As its name implies, the software was designed to determine what kind of damage (a hole, or crater) would be put into a shuttle silica tile when struck on the way to orbit. It's important to note here that the Crater program was designed to analyze damage to the silica tile, and not the RCC where it's now assumed the most critical damage took place. Crater was also built using results from tile impact tests run years ago, and the impactor used as a basis for Crater was a piece of foam approximately 400 times smaller than what was estimated to strike the orbiter. Through all these assumptions and difficulties, the board noted that Crater's results did a respectable job at characterizing the damage. In fact, Crater predicted that the foam strike would have even cut through the orbiter's silica tile and exposed *Columbia*'s skin. This result was dismissed because Crater is known to be conservative, estimating more damage than was observed in the past.

By flight day four (Sunday, January 19th), the Crater analysis was completed, showing that it was quite possible that the orbiter was severely damaged by the foam strike. Rodney Rocha, working with the analysis, e-mailed the engineering directorate to see if they were preparing a request for the crew to check the wing for damage from the flight deck. He never received an answer, and didn't follow up to get an answer. The board listed this as Missed Opportunity One.

On Tuesday morning, the 21st of January (Monday, the 20th of January was the Martin Luther King Jr. holiday, so no meetings were held), meetings began in earnest to discuss the mission in general, and the foam strike in particular. The board noted that the 21st of January was the first day that the Mission Management Team meeting took place, even though mission rules called for the meeting to take place every day that there was a mission underway. At that meeting, the debris strike was discussed and compared to previous instances of foam impacts. In transcripts of the meeting, mission managers, who were also scheduled to serve as mission managers for upcoming missions, were thinking more about the effect that the foam strike would have on later missions, not *Columbia* in flight. As was stated several times in the transcripts, when on-orbit damage is brought up, the manager in question says "there is not much we can do about it." The problem here comes into play because managers in charge of the current mission are making decisions based on future missions, not thinking first of the safety of the crew in flight. The only time safety is even brought up is when the foam strike is compared to previous missions that experienced foam strikes and returned, so they rationalized that it would be a maintenance problem for this mission, not a crew safety issue.

It's here that the board placed missed opportunity two, though it sort of pervades the rest of the mission. As a routine, the orbiter crew takes pictures of the external tank as it separates. This was done during *Columbia*'s mission, and a portion of the images taken was beamed back to Earth for analysis. Unfortunately, the portion of the film beamed down did not show the area of the tank where foam was suspected of coming off. All indications are that other footage existed, as the normal procedure produced useful images of the bipod area in the past. No one asked for the additional footage to be sent down.

At this point, in discussions and e-mails, the management position that the strike was no big deal, just a maintenance issue that would have to be dealt with upon landing, began to spread through formal and informal chains of command. Formal chains of command are between people who are associated within an organization (mission operations, or mission engineering, for example) while informal chains of command allow one organization to influence the other, perhaps at the expense of a rigorous analysis of the problem by the influenced organization. One particular e-mail, only one line long, stated that the impact "should not be a problem" from a person known throughout shuttle management as an expert on thermal protection. While this person is an expert on the tiles, a portion of the thermal protection system (TPS), he is not an expert on the RCC panels where damage was suspected, and no one questioned his knowledge. Therefore, though his statement was likely correct from his point of view (the board says it's likely that tiles were not significantly damaged in the strike) it was incorrect overall (the board found it likely that the RCC panels were damaged in the strike).

Another awkward situation arose here, contributing to management convincing themselves that the foam was not a problem. This was not the first time foam was lost during liftoff, and previous experience with it proved that missions could return after foam strikes. A particularly bad strike on a recent mission (STS-112, two missions before *Columbia*) had been classified as much less of an issue than previous ones in post-flight reviews, and managers had briefing charts stating that foam loss was never a safety-of-flight issue, just a maintenance issue.

At an unrelated meeting between NASA and the National Imagery and Mapping Agency (NIMA) in Washington, DC, NASA personnel mentioned the foam strike. Gathering imagery of the shuttle on orbit was discussed, but no action was taken. The board described this as missed opportunity 3.

Two other efforts started on this day for imagery of *Columbia*. One took place informally, where the manager of the Debris Assessment Team for USA (The United Space Alliance, a contractor organization charged with operating the shuttle) called a NASA manager, asking what would be required to make such a request. The NASA manager contacted the DoD Manned Spaceflight Support Office, and the board notes that this was not the proper procedure. The other effort started with a more formal mandate, but it took place through the engineering directorate, instead of the mission directorate. The board points out that a mission directorate request would have carried more weight.

Missed opportunity four took place on flight day 7, Wednesday, and involves another manager starting a process to request imagery of the orbiter. Once again, this was not through official channels, although action did start.

Missed opportunities five and six were combined because they refer to NASA's safety office (in the formal organization, the safety office is separate from the mission organization, because it is supposed to give them a clarity of looking at things without other concerns like mission timelines and space station construction) being informed of the strike. The two safety managers contacted were from Johnson Spaceflight Center and from headquarters, and they were told that the foam strike was an "in family" event, meaning that strikes like it were seen before with little consequence. The missed opportunities arose in that neither of the men briefed pressed the issue of imaging the orbiter, with one of them actually stating that he would defer to shuttle management on the request. Both actions are counter to how a safety office should work. If anything, a safety office should go overboard to verify that an event didn't have safety ramifications.

At 8:30 on Wednesday, a call was made to the Air Force to cancel on-orbit imaging of *Columbia*. The reasons cited were that NASA had the capabilities to do the required assessment in-house, and didn't require DoD's assistance. It appears as though the Mission Management Team Chair cancelled the request after polling those who'd informally requested the data for the requirement making imagery necessary. Here, the people who'd requested the data were stuck, because they wanted more information about the current state of *Columbia*, but didn't have a firm "requirement" for it. The word requirement can have a deep meaning in bureaucratic organizations, specifying formally why something needs to be done. After pressing for requirements and receiving none, the Mission Management Team Chair cancelled the imaging request, also citing that interrupting flight operations to point *Columbia* properly (in the correct direction so that images could be taken) would cause a mission impact.

The board noted that no one in the operational chain of command, making decisions about imaging *Columbia*, understood the full capabilities of the DoD, because none of them had the proper clearances. This allowed further rationalization against taking images since those making the decisions believed that the images wouldn't be very good.

Even after "turning off" the imagery request, the Mission Management Team Chair sent e-mails asking questions about the foam strike, and whether or not there could be a "safe size" of foam that could be ruled out as causing a safety-of-flight issue. Here, I believe the goal was being able to answer a question at a press conference. Answers received stated that a general statement could not be made—there was no way to state that one piece of foam would cause damage while another, smaller one would not.

In preparation for another meeting of the Debris Assessment Team, members tried to interpret the (in some cases rumored) cancellation of their imagery request. Discussion turned to whether they had a "mandatory need" for imagery. When the debris team asked mission managers what a "mandatory need" was, they got no answer. This meant that Debris Assessment Team members had to make a case for damage to the wing without the data (the images) that they needed to make the case.

After this meeting, one manager typed an e-mail talking about NASA's response to the foam assessment using words like "bordering on irresponsible" and "potentially grave hazards," but did not send it. When asked why, he stated that he'd already brought up his concerns, and sending out such an e-mail would go outside his chain of command.

In a darkly humorous note about computer systems within NASA, a note in the mission log states that someone trying to view the foam impact footage doesn't have the proper software, and needed another format or a VHS tape. This proves that even NASA hasn't mastered the concept of standardized computer software.

Flight day eight, Thursday, January 23rd, dawned with two NASA managers discussing what would have been feasible for imaging if DoD assets were invoked. Neither of the managers had the information (clearances to know about imaging capabilities) to make the discussion useful, and neither pressed for more information about capabilities or re-opened a request for imagery. This is categorized as missed opportunity seven.

Another Debris Assessment Team meeting took place on flight day eight, and in preparation for the meeting, the team debated putting a slide in their presentation asking again for imagery of *Columbia*. The slide was not included. Alternate return profiles were discussed that could protect the possibly damaged area of the wing from hot gasses (there were none), and some members of the team asked why they were still analyzing the strike if management already decided that it wasn't an issue. They were told that it was good training for new analysts.

Flight day nine started with a briefing from USA to the Mission Evaluation Room manager about the Debris Assessment Team's findings. This meeting was filled beyond capacity, with people standing out in the hallway. In the briefing, the final assessment came down that there was no safety-of-flight issue, but that there were a number of uncertainties in their data. Uncertainties included the size of the foam and where the strike took place. Engineers who attended the meeting felt that the manger dwelled on the answer (no safety-of-flight) instead of the concerns about the answer. This manager, in turn, briefed the results of the analysis to the Mission Management Team, and no technical questions were asked. Some transcripted discussion, quoted in the board's report, shows the focus that the Mission Management Team has on turnaround issues for the next flight, instead of the current flight in progress.

Over the next seven days, the Debris Assessment Team continued to work on scenarios as to what could be damaged based on different strike locations for the foam. NASA management got word of this ongoing investigation, and called the contractor involved to express concern that work continued. In response, the contractor restated their confidence in their analysis, and the NASA manager in question sent an e-mail to make sure that the Mission Management Team understood that the issue was not closed, but that it still didn't look like a problem.

Analysis continued at a low level, and eventually word got to Langley Research Center in Virginia that tile damage (or damage to the left landing gear door) might destroy the left main landing gear before landing. Engineers there took it upon themselves as an informal tasking to determine whether such a landing would be survivable, or if the crew would have to bail out and let the orbiter ditch into the ocean. Because there wasn't an official task assignment, these engineers felt that they couldn't move astronauts out of their normal training rotations in the simulator, and instead worked after hours to create a two-flat-tire landing scenario. Their scenario showed that the landing would be survivable, but very serious. Three days before *Columbia* was scheduled to return, an e-mail went from Langley Research Center to Johnson Space Center asking whether there was "any more activity today on the tile damage or are people just relegated to crossing their fingers and hoping for the best?"

It didn't matter. *Columbia*'s fate was sealed. On Saturday morning, February 1st, the crew prepared for a normal return that was anything but.

Underlying Causes

The Columbia Accident Investigation Board went to great lengths to track down, as closely as they could, the cause for *Columbia*'s breakup over Texas in early 2003. Examining data, debris, and maintenance records provided a pretty clear picture as to the mechanical failures that, when taken in the proper sequence, caused the tragedy. The panel dug deeper, though, trying to find whether the *Columbia* disaster was a random chance accident, something that, though tragic, was unforeseeable on this particular mission similar to an aircraft crash, or was it simply the natural result of changing times in a government-run space program. The answer, it turns out, is much closer to the latter than the former.

While the report tracks the progression of the US Space Program from the *Challenger* explosion to the *Columbia* breakup, the roots for both events actually run deeper, into the very founding of the National Aeronautics and Space Administration. NASA was founded during the cold war as a non-warfare tool in the propaganda war between the United States and the Soviet Union. As mentioned earlier, first successes by the Soviets required action on our part, and the founding of NASA was part of that action. A few years after NASA was founded, it was given the most audacious goal ever imagined: land a man on the moon and return him safely before the end of the decade. Because of NASA's high priority, being a non-lethal weapon of the cold war, this goal was seen as a national priority. Funding levels, though not completely stable and not guaranteed, were high and congressional support was strong. Eventually, after a long, rough road, the goal was achieved in 1969. But what then?

The war in Vietnam became increasing news as the space program reached its peak. After *Apollo*'s landing on the moon, détente became the operative word between the US and the USSR, with the space program even playing a role with the 1975 Apollo-Soyuz mission linking the superpowers in orbit. Through the 1980s, as the space shuttle started flying, space played less of a role in the superpowers showdown, with the soviet space station Mir flying while the space shuttle made repeated trips into space. President Reagan proposed the Space Station Freedom, but as time went on and the station's budget grew, its future was questionable.

Then, in 1990, the cold war abruptly ended. The founding reason for having a space program, national prestige through peaceful competition, was gone. Even though the political climate that surrounded the US space program changed dramatically, the space program itself did not. With a national priority much lower than its previous levels, NASA found itself battling for funds with other government organizations. For years, NASA's budget remained flat, not even adjusted for inflation. The increased fiscal pressure, combined with several mandated hiring freezes (no new NASA employees could be hired) and decreasing luster in government, conspired to make conditions within the agency difficult. These difficulties manifested themselves in several ways.

Management Reforms

During the 1990s, management reform was the buzz word throughout the business world, and that philosophy carried over into NASA. Taking several guises, such as Faster-Better-Cheaper, International Organization for Standardization (ISO), and insight vs. oversight, these alternatives were designed to cut costs and make activities more efficient.

Faster-Better-Cheaper was the approach that caught the most press during the decade, probably because of its appeal to the general public: if space travel could be made cheaper, we could do more of it. The idea came about during Dan Goldin's watch as head of NASA, and was at least partially inspired by the very public failure of the *Mars Observer* spacecraft. *Mars Observer* was a large orbiter, designed to fly to Mars in the early '90s and return images and other data of the surface. Ground lost contact with the craft just as it was approaching Mars, and an investigation showed that the likely cause was an explosion within the propulsion system. Because *Mars Observer* was such a large and expensive mission, when the accident happened there were no other serious missions to Mars in the planning phases. The announcement of possible microfossils found in the meteorite from Mars placed the exploration spotlight directly on the red planet, and the response was a plan for two missions to Mars each time the opportunity presented itself. This could only be accomplished if the missions were much smaller in size and budget, and accomplished smaller goals on each mission. Under this plan, if one mission were lost, there would not be a huge gap in data, and the instruments designed to fly on the lost mission could eventually be flown on others. In theory, the approach sounded fantastic.

FBC started off with a string of successes. The *Mars Pathfinder* lander and *Mars Global Surveyor* orbiter missions of 1997 were huge hits, both publicly and scientifically (MGS was much larger scientifically than *Pathfinder*, in fact it ranks 8th in a list of scientific accomplishments of NASA and is still functioning today). Another program under FBC, called the Discovery Missions, allowed researchers from outside NASA to propose spacecraft accomplishing different goals in the solar system. The first Discovery Mission, called Near Earth Asteroid Rendezvous or NEAR, was launched in the same timeframe of *Mars Pathfinder* and its sister orbiter. NEAR was designed to orbit an asteroid, which it eventually did after some additional flight time due to a thruster malfunction. NEAR returned tremendous images of the asteroid Eros, circling ever closer to the spinning rock, and eventually, on February 12th, 2001, NEAR actually landed on its quarry, the first asteroid landing in history.

FBC and its early successes sent shockwaves through NASA, forcing most segments of the agency to look at their methods of doing business to see if they could benefit from the approach. Even a program that I work with, the Geostationary Operational Environmental Satellite (GOES, which is actually a National Oceanographic and Atmospheric Administration [NOAA] program that NASA purchases the satellites for) worked to purchase its next round of satellites much more cheaply, though contractor turnover (discussed later) also played a part in the efforts to lower cost.

I think that FBC in its initial concept was a great idea. Having worked on space programs, however, I understand that the easiest of the three components to measure is cheaper. I assume that this fact is the same for business interests as well. Because it's the easiest to measure, it's also the easiest to audit, and the easiest to give managers grief over for not meeting cost goals. With people focusing on money, the push to cut costs rose. The missions were originally envisioned to carry one or two major instruments on board, but pressure rose to do more and more with the smaller budgets. The human spaceflight portion of NASA's budget was also subject to the same cost-cutting pressure, although not in the same form.

The International Standards Organization, founders of ISO 9001, and its earlier form ISO 9000, created business practices designed to document processes across a company. The core idea is that a properly documented process is one that anyone within the company can take over. This would prevent single points of failure in people who, if they suddenly left the company, would impact production. The accident investigation board points out that these ISO practices are designed for manufacturing environment, which space operations surely is not right now. As we'll get in to later, about the only way space travel will ever become cheap enough for its full utilization is for it to become an industrial process. That is, repetitive on the order of hours (even minutes) instead of there being months between missions. At that time, a system like ISO may be applicable, but the practices are not easily applied to today's activities in space.

Insight vs. Oversight, another management reform, also swept the Air Force launch business during the 1990s. This idea was based on increasing commercial space activities, and that the company's need to compete for commercial contracts would keep its practices sharp and maintain a high mission success rate in their launch vehicles. In NASA's crewed spaceflight area, mission success translates directly into crew safety. Because of this competitive force keeping safety standards high, NASA (and the Air Force) could afford to scale back their own involvement in space activities. An example of this type of practice is that, under the oversight system, the Air Force would have a military person participating in a review of some sort of anomalous event. This Air Force person would have worked in the area, and would know if the company was on the right track in the review. They would have a say as to whether or not the review was satisfactorily completed. Insight, on the other hand, has the contractor carry out their investigation, briefing the final results to Air Force (or NASA) management. In a perfect world, the manager being briefed would know enough about the problem to ascertain that a proper review was done, and that if this weren't the case, independent reviewers such as the NASA safety office would question results and analysis if necessary. As described above in NASA's response to the foam-strike analysis that took place during the *Columbia* mission, neither situation happened. Management focused on the conclusion "no safety-of-flight issue," which coincided with what they wanted to hear, and didn't question whether all possible actions were taken. The reason they wanted to hear that the foam strike didn't pose a safety-of-flight issue was the space station completion schedule.

Schedule pressure was another factor cited in the *Columbia* report as a contributing factor to the tragedy. Ironically, the schedule pressure did not come from some sort of mandate similar to our Apollo moon landing goal, but came about due to budgets and a rather strange coincidence of personnel change within NASA leadership.

In 2001, after President Bush took office, the space community waited with bated breath to see who would be named as the new NASA administrator. Dan Goldin, who'd served in the position since the first Bush Presidency, had held the position for longer than anyone in the agency's history. Several names drifted across the rumor radar, but no one seemed interested in the job.

Compounding curiosity about the upcoming choice was NASA's recent admission that the International Space Station budget had once again risen, and this time the dollar amount exceeded the dollar figure allowed by Congress. The newly elected President assigned a member of his accounting office (the Office of Management and Budget, or OMB) to look into NASA's dealings and come up with a plan to resolve the station problem. That OMB member was Sean O'Keefe. While investigating NASA's accounting practices, he found that there were no standardized practices across the agency. The differences in approaches led to problems that could easily cause bad estimates and lead to cost overruns like the one recently reported for ISS. Mr. O'Keefe came up with a plan: give NASA a preliminary goal, short of the entire station completion, and a deadline to meet it by. If that goal was met on time and on budget, then the administration would review completing the station with additional funds. This plan was approved by the President, and hit the headlines in the form of canceling large portions of the station, including the habitation module and the crew return vehicle. These items were required to bring the space station crew up to the planned 7. Without the habitation module or the crew return vehicle, the maximum number of people who could be left on board the station without an additional ferry craft was three. The cuts angered some international partners, who counted on the larger capacity to get their own astronauts on board. The goal was set to reach the "US Core Complete" stage of construction by February 19, 2004. The station was safe for the time being, however, and work continued on the outpost in space.

Soon after this plan was approved, word came out that a new NASA administrator had been found, and his name was Sean O'Keefe. So now the person who'd come up with the plan for space station construction meeting goals and budgets was in charge of building the station. Mr. O'Keefe set immediately to work on standardizing NASA's accounting practices, and set all eyes on the goal for US Core Complete.

There's a military saying that no plan survives first contact with the enemy, and the original, methodical space station plan did not survive long-term contact with reality. While the schedule to reach US Core complete was reasonable when the plan was put into motion, several events, some inside and some outside NASA's control, made the schedule less and less realistic. Maintenance problems with the entire space shuttle fleet grounded the craft for more than a month on more than one occasion. The maintenance problems put additional pressure on the schedule because some orbiters were scheduled for periodic maintenance and upgrades, and delays in their missions meant that the maintenance had to be delayed or rescheduled. As normal operations delays happened to individual missions, schedules got squeezed even further, forcing major policy changes, such as refitting the orbiter *Columbia* to fly to the ISS. Original policy held that *Columbia*, the oldest and heaviest of the orbiters, could not take enough useful cargo to ISS to make the necessary upgrades cost-effective. Maintenance schedules on other orbiters, however, drove the decision to make *Columbia* ISS-compatible and allow it to fly to the station one time, making a relatively low-weight delivery.

Even with the *Columbia* refit plan, schedules grew tighter, and NASA management wanted monthly updates describing progress towards the February 2004 goal. In case the constant requests for status was not enough to show how important upper management felt the deadline was, NASA workers were given a screen saver for their computers, counting down the days (and separately counting the total minutes and total seconds) to the imposed deadline for US Core Complete of ISS. Workers interviewed for the report described the screen saver as something amusing, but they knew that the deadline was not feasible. NASA management saw it as feasible and squeezed every known (and some made up) space in the schedule for more time. Future flights, not even in the air yet, were having their operations affected, because briefing slides in the end of 2002 showed that there was no room in schedules for an orbiter to land at Edwards Air Force Base anymore. This option, kept open in case of foul weather at the primary landing site in Florida, would cost an extra week in ferrying the orbiter across the country for its next mission. Charts showed a labor force ramp-up, and people already working would be pressed into service over weekends and holidays, all to meet the deadline.

When interviewed, NASA managers said that the schedule pressure did not force unsafe or unsound decisions related to operations or processing. Workers on the floor said that schedule pressures did lead to unsafe or unsound decisions, but that they felt powerless to claim that as the case, because there was no failure to point to as a symptom.

February 19, 2004, passed without the ISS achieving US Core Complete status.

Contractor Turnover (lack of independent viewpoints)

One effort that took place to try and cut the costs of shuttle operations was to turn an increasing role over to contractors. It was thought that contractors, or corporate interests, have a more efficient structure that is unfettered by Federal policies and practices. The theory went that if a company were simply given the task to run the space shuttle safely, they would eventually settle upon the cheapest way to do that through experience. A partnership of two typical rivals, Boeing and Lockheed Martin, came together and named themselves USA (United Space Alliance) winning the contract to take over space shuttle operations. NASA held on to the core safety role, since in theory they would be immune to the profit motive. As we've seen, NASA was not immune to the schedule motive.

Some problems came up early and remain today with this transfer. The human nature part of the problem is that no entity, government or otherwise, likes handing over its role (read as jobs) to another entity. Anyone who used to do job X who now has to either switch to a company to do job X or move to a different role watching others do job X is not going to be happy. Enough people not being happy about a change, along with their supervisors and managers, can work to slow or even stop a process.

Another problem comes in that the Space Transportation System was designed to optimize as much as possible during the construction phase, but had very little money spent to make operations and maintenance easier. In many cases, procedures carried out on the shuttle as part of their turnaround procedure were the only way to do certain things. While some efficiency could be realized through changes, these were pretty limited. I believe that this is a symptom of a corporate entity taking over for a government-procured system. If the corporate entity was given the freedom to build the craft that they wanted, and then paid to operate it, the savings related to operations and maintenance would be much greater.

The turnover to USA was taking place, however, and this served to further dilute NASA's knowledge of the systems that they were charged to keep safe. Managers with less technical knowledge were forced to take the analysis of corporate experts, and lacked the technical knowledge to ask useful questions. The corporate analyzers (and their managers) were put in the difficult position of deciding whether or not the hardware they were charged to fly (in Boeing's case, a subsidiary actually build the space shuttle orbiters) was safe to continue to fly. Now, if you're a corporate entity, and your future contract status with NASA is riding on whether or not the customer is happy, it's almost impossible to avoid painting a rosier picture of any situation. In all cases before *Columbia*, watering-down bad news did not lead to any serious effects. In late January of 2003, this approach met its limit, and resulted in a dark February 1st.

We've discussed the changing roles that NASA played in national security and prestige, but so far have not talked much about how those changing roles affected the organization itself. NASA is not the agency it was in the 1960s and early '70s. The lack of a national goal (the race to the moon) and a gradual, grinding decrease in funding wore the agency down. Because of a lack of a coherent, driving goal with a timeline, anyone at NASA with a manager's ear could create a research project based on the idea that it could be used, one day, for such a goal when it came about. Robert Zubrin compares this method of research to a couple who want to build a house sometime. While waiting, they ask what friends did in building their house, and purchase related items for their own house. When the time comes to build it, the couple asks an architect to design a house using all the parts that they purchased. Some of the parts they bought will work together, but some of them will be drastically wrong. Building a house using all of them will be a very expensive endeavor. On the other hand, the couple that wants to build a house who hires an architect from the start is on a better track. The architect draws up the plans and the couple purchases the parts to implement that plan. The house is buildable and reflects what the couple wants and can afford.

Even if a major project doesn't come along, as exampled by the house above, this pick-and-choose method of research doesn't work well in keeping an agency vibrant. When funding arguments come, and within government programs they are as predictable as the seasons, each one of these projects had to be argued based on their potential merits of fitting into a plan that doesn't exist. Without any real timeline driving the projects, some of them get cancelled while most of them simply get extended to save money in the current year at the cost of future years. Over these years, more ideas come up that get funded and thrown into the pile for next year's funding debates.

Crewed spaceflight faced something similar, though not as much money is devoted to research here. In this arena, the money sink was development of the ISS. Over time, as costs swelled, managers sought funding from other areas. The easiest area for them to access was shuttle operations, since it was the closest to them organizationally. Later, to make such transfers even easier, shuttle and ISS monies were all lumped together. As the ISS budget grew, safety upgrades to the shuttle fleet were delayed, again saving money today but driving up overall costs.

While the breakup of the *Columbia* over Texas was shocking to many, the Columbia Accident Investigation Board found that some form of accident was inevitable. It became that way given decaying practices over time with human spaceflight. The board came up with recommendations to fix things, however, and included them in their report.

Recommendations

The board recommendations were formally divided into two parts, but in my opinion, there was a third part that dealt with setting national goals for space. The two formal parts were assigned different priorities. The first group was the return-to-flight recommendations, that is, these actions must take place before the shuttle flies again.

The first couple recommendations described here (they're in an order that fits with the text within the report) have to do with the foam impact directly tied to the loss of *Columbia*. In them, NASA must take time to understand the causes for external tank foam shedding and make efforts to stop it. In case the complete stoppage of foam shedding is impossible, the space agency must completely quantify the damage that shedding foam can cause to the shuttle stack in all possible strike areas. Presumably, they will implement this portion of the recommendations after making the orbiter more resistant to minor strikes on ascent, another return-to-flight requirement. To provide nearly instantaneous feedback on whether or not such an event occurs, the board requires NASA to have no fewer than three cameras on the ground trained on the orbiter during launch through separation of the solid rocket boosters. While it's unstated, I assume that the board also means that those cameras must be in focus, as the primary camera focused on Columbia the day of its final launch was out of focus.

Imaging requirements follow the orbiter into orbit. When the external tank separates from the spaceplane on each mission, NASA must find a way to take multiple, high-resolution images of the tank and beam them to Earth. A detailed survey is also required of the wing leading edges upon arrival to orbit. Also, to avoid future arguments over whether satellite images of the orbiter are required, the board recommended that imaging an orbiter during flight become a standard requirement.

In case all the recommended safeguards do not prevent a future compromise of the orbiter's thermal protection system, the orbiter must be able to diagnose and fix any portion of the system while on orbit. There was some speculation that the idea of completely unaided repair wouldn't be necessary in the case of a mission to the ISS, but the board wisely brought up cases where the orbiter was due to rendezvous with the ISS, but was unable to due to some problem. Also, there's the possibility that the orbiter would get damaged separating from the station, in which case the crew would have to repair the damage for themselves.

In processing between missions, at least two employees must be present when an area of the shuttle is closed out before flight, and whenever some sort of hand-done maintenance is required. The region on the External Tank where foam fell off is processed by hand because of its complex geometry—most other areas on the tank are covered in insulation by machine.

The board also recommended that NASA stop using confusing terms such as "processing debris," which was a category created for debris damage found during preparation for launch that only caused maintenance problems. NASA is now required to adopt the more stringent (but standardized) aeronautical term of "Foreign Object Debris" for all events that involve debris. This will make each debris incident more serious, and prevent a sidelining of some events. In another modernization effort, the board asked that documentation showing shuttle changes since the design phase be updated and digitized. Also, the photographs taken as part of "closeout" operations, that is, when a system is considered ready for flight and shut tight, are now to be digitized and immediately available while the mission is flying. A seemingly unrelated recommendation, though it was considered as one possibility for a contributing cause of *Columbia*'s breakup, is a requirement to test and qualify bolt catchers that are used to keep small pieces of the shuttle stack from flying apart and damaging the craft when large pieces are supposed to separate.

The last three recommendations move away from hardware and safety of individual flight issues and into the realm of the decision-making process before and during each flight. In them, the board recommends that NASA maintain a flight rate for which it has the resources to keep up, and that the agency doesn't try to push things harder than they should. I believe that this will be the toughest of all recommendations to follow in the long term. Mechanical fixes are easy compared to the task of maintaining a focus for the years remaining in the shuttle program. Of course, this same recommendation should be applied to any future spacecraft that NASA flies as well.

In an effort to build up an independent engineering and technical branch responsible for all requirements (discussed in the coming non-RTF section), the board requires that NASA draft a plan for doing so before the shuttle returns to flight. The board also recommends a training program for shuttle managers that forces them to make decisions like they should have made during *Columbia*'s mission, using all the resources at their disposal as if a shuttle mission were flying. The management team would be presented with partial information and asked to make a decision based on additional data they could gather through support agencies, like the debris assessment team that played a role in *Columbia*.

The non-return-to-flight items specified in the report are designed to improve overall shuttle safety, even though they may not have made any difference in the STS-107. Another portion of the recommendations deals with the idea of bringing the documentation of the system up to modern-day standards. In many cases these recommendations are prefaced with the phrase "to the extent possible," giving NASA some leeway on their implementation.

The first batch of recommendations have to deal with the thermal protection system on board the shuttle, asking NASA to work on improving the shuttle's ability to reenter Earth's atmosphere with minor damage to the leading edge of the wing. Supposedly, the damage sustained by *Columbia*, according to tests, would not qualify as minor damage. The board also asked that NASA work to understand the properties of RCC better, so that analysis of any damage to the material will have a stronger base to it. Launch pad maintenance should also improve so that RCC components are not exposed to zinc (a material that can affect RCC's strength and/or ability to withstand heat). In case all of this additional analysis causes more RCC panels to be replaced, the board recommends that NASA purchase a whole raft of supplies of the panels. That way, replacement for an RCC panel will not cause a delay in processing between missions.

To improve the analysis capabilities for the debris assessment team, the board recommended that NASA develop, validate, and maintain a physics-based analysis tool to evaluate debris damage to the thermal protection system. This would provide an analysis capability similar to what the board had in its investigation. Using this new software, the board recommends that experts set damage (as estimated by the software) thresholds above which certain actions will be taken, such as inspection or repair to the affected system. Setting such limits now, while a crew's life is not on the line, cuts down debate when the numbers are important.

The board felt that the Modular Auxiliary Data System (MADS) provided so much data as part of their investigation they recommended that the system be upgraded and expanded. Upgraded to use the latest sensor technologies, and expanded to allow data from the MADS records to be beamed back to Earth in real time. In the case of *Columbia*, the data on the MADS system could have been used to show damage to the left wing without any imaging or effort on the pat of the crew, because one MADS sensor showed a slightly different temperature profile than was normal for a launch.

The wiring on board the orbiters has the potential for being used for up to forty years, and the board asked that NASA develop a method to inspect all of it.

To try and cut the chance of an orbital collision with space junk causing the loss of an orbiter, NASA is required to move towards applying the same practices used for the ISS to a shuttle mission in progress. Currently, the space station is given a wider berth than a shuttle between it and any space debris that might collide with it.

In some sweeping recommendations for NASA organization the board recommended that an independent technical and engineering authority be set up to monitor standards, waivers, and hazards, and to independently certify launch readiness before each mission. They also asked that the office of safety and mission assurance be combined into one unit overseeing the entire space shuttle program. Both organizations are to be funded directly from headquarters, and are to be free of schedule pressures and concerns about program costs. By funding them independently of the shuttle program, they are shielded from budget cuts within the shuttle program, because safety and mission assurance were some of the first areas cut in the past.

Like the fragmented safety office, the shuttle integration office is spread across field centers, making overall integration of the orbiter, external tank, and solid rocket boosters difficult. The board recommends that this office be combined to include all portions of the shuttle stack.

Eyeing the future, the final two recommendations deal with operating the space shuttle past the year 2010. In the first, the board asks that NASA develop a re-certification process and inspect the shuttle system to its core before operating the vehicle past the year 2010. As a method of making this effort easier, NASA's been asked to review all shuttle drawings for accuracy, and convert them to a computer-aided design format to allow simpler access.

Finally, after detailing the mechanical and organizational problems that caused the loss of the *Columbia*, the Columbia Accident Investigation Board asked that the United States take a look at what it's doing in space. While many of the organizational problems within NASA could easily be written off as a natural course of events for a bureaucratic organization over time, the CAIB went deeper and pointed out that NASA's position within the United States' priorities had shifted. This shift, away from the front lines of the Cold War, pressed on an organization used to success and wanting to maintain a critical role in the country's psyche. To solve this problem, the CAIB called for a national debate on space, to decide whether we should continue to explore and exploit it, or we should turn away. If we choose to continue, then we need to decide the level at which we'll continue, set funding to meet that priority, and press on.

The Road Back

Word leaked out before the official report release date that it would be harsh on NASA's practices and organization. NASA stated a position early on that it would implement the board's recommendations and more in response to the loss of *Columbia*. They even asked for preliminary recommendations so that they could address them early on. In the meantime, the agency had an international project on orbit, the ISS, with three (later two) people on board that had to be taken care of.

First and foremost, NASA had to respond to the day-to-day needs of the crew aboard the ISS. The space shuttle is critical to assembly procedures, in that nearly all of the new equipment brought to the station is carried up in an orbiter cargo bay. It was clear that space station assembly would have to stop until the shuttle was flying again. It was also determined that, without the shuttle making trips to the station, the outpost could not support 3 people—its normal crew. The plan was put into place where a caretaker crew of two would live aboard the station until regular shuttle flights started again. A crew of two can survive on the normal cargo deliveries made by Russian supply craft, and would maintain the goal of keeping humans in space full-time since the beginning of the space station program. Normal crew rotations would take place using Russian *Soyuz* capsules to ferry the new crew up. *Soyuz* capsules must be replaced about every six months, and the crew on board the station takes theirs back to Earth after the new one docks. This system would work as long as there were no major failures on board the station, which would require a shuttle trip to bring up the replacement parts. Any crew that was present in case of such a failure would simply climb aboard their *Soyuz* craft and return home. It was hoped that shuttle return-to-flight would happen rather quickly, and that the caretaker status of the station would not last too long. As downtime surpassed year, now looking to be more like two years, with the first shuttle flight expected to be a demo of new procedures (more on that in a bit), NASA may have to rely more on foreign partners and their launch capability. Besides the Russian *Progress* cargo ship, the Europeans are building the Automated Transfer Vehicle (ATV) to carry supplies as well. The ATV will be able to carry much more payload to the station than the *Progress* vehicles, though exact plans of what the ATV's impact will be on the station have not been described.

As this book went to press, the system was working, with Gennady Padalka and Mike Finke on board the station for Expedition 9, and Leroy Chiao and Salizhan Sharipov ready for liftoff on a glitch-delayed Expedition 10. The station is not without its problems, however, with control systems' popped circuit breakers causing a need for unplanned spacewalks and a faulty oxygen generator requiring the use of several backup plans. It remains to be seen whether or not the station can continue to function during the shuttle hiatus, or if the crew will have to use their *Soyuz* craft to leave the ISS. Abandoning the station would be risky to the entire project, however, because it's possible that, once abandoned, the station could never be re-inhabited. The ISS is designed with on-board maintenance in mind, because crews would always be there to fix anything that went wrong. If something were to drive the crew away, failures within the station could cause it to tumble or become uninhabitable. While no dates are mentioned in plans related to the station, this kind of pressure, keeping the station crewed to keep the program alive, sounds remarkably similar to schedule pressures that caused the *Columbia* disaster, as reported by the CAIB.

In preparing for shuttle return-to-flight, NASA has created a web page to chronicle their plans and progress. The website includes a downloadable document that summarizes the effort to date[40].

In the document, NASA personnel list every CAIB recommendation and then show the plan to address it. Return-to-flight items are highlighted, and schedules provide insight into when the space shuttle is likely to fly again. The document is described as a living document, to be updated as conditions warrant. The copy I downloaded in January of 2004 was dated November 2003. NASA also includes actions that they're taking above and beyond the CAIB recommendations. These actions are meant to show the agency's ability to diagnose its own ills, and show commitment to fixing underlying problems that perhaps the CAIB missed. Most of them don't relate directly to the *Columbia* accident (this makes sense, since if the changes did relate directly, the CAIB would have made them recommendations), and many deal with ground procedures that should plug the holes that decisions leaked through during *Columbia*'s final flight. Some examples of these areas include: evaluating the risk to the public of shuttle overflight, quantifying risks, and setting acceptable levels for risk, and formalizing the ideas of waivers, deviations, and exceptions used for when something does not meet the necessary standard.

The press has focused on the inspection and repair plans for the shuttle, likely because they provide the best visuals and relate so closely to the breakup in February 2003. There have been several news articles about the shuttle arm, which will now be required on all missions. When an orbiter reaches orbit, the arm will pick up an extension that has a camera on the far end. Once grappled this extension will be maneuvered to give the crew a complete view of the exterior of the orbiter's thermal protection system. This initial inspection is expected to take place in the first orbits of a mission and provide the crew with assurance that their vehicle handled the flight into orbit well and will reasonably support their return from space. If a problem spot is discovered, astronauts are currently training on methods to repair that spot and allow a safe return.

This type of capability would be critical in case the shuttle were planning to fly in an orbit where it could not dock with the ISS, similar to the orbit that *Columbia* was in on its last mission. The only outstanding mission that required such flight was the Hubble Space Telescope servicing mission #4, where an orbiter (originally scheduled to be *Columbia*) would fly to the orbiting telescope and replace life-limited items and a few instruments for the last time. Two days after President Bush's announcement to re-invigorate NASA, however, the final Hubble mission was cancelled. The official story describes the timing as coincidence, with the servicing mission cancellation not feeding directly into the new space efforts. Statements go further to say that shuttle missions will be devoted to the ISS, and that safety reasons prevent a dedicated mission to Hubble. News reports describe an effort to build an uncrewed vehicle, designed to fly to Hubble, attach to the telescope, and bring it back to a fiery end in Earth's atmosphere, but official details are sketchy.

The cancellation of the last servicing mission, derided in the press as the decommissioning of Hubble, has lead to a groundswell of scientific and layperson interest in saving the orbiting telescope. Barbara Mikulski, a senator from Maryland and long-time supporter of the Hubble Space Telescope (not coincidently because the telescope is controlled out of her home state), came to the observatory's defense. She wrote a letter to NASA administrator O'Keefe asking that he reconsider the decision. The NASA administrator said that he'd ask Admiral Gehman for an opinion, and Senator Mikulski went on the record saying that she would accept Admiral Gehman's recommendation. Admiral Gehman responded with a letter basically stating that he couldn't make a firm recommendation without reforming the Columbia Accident Investigation Board, but that either interpretation (NASA's decision not to fly, or protester's desire to fly) would be correct. He described a Hubble-servicing mission as "slightly more risky" than an ISS delivery mission, and said that a thorough analysis of the risks and benefits of the mission was required. Critics have called this a weak position to take.

In a debate with Robert Park about the merits of human spaceflight vs. exploring with robotic craft, Robert Zubrin went on record saying that he believed a mission to HST was very important, for two reasons. First of all,

Hubble is the greatest scientific achievement in NASA's history. Adding one shuttle mission to an expected manifest of 25 flights to complete the space station would add minimal risk to the overall program, and would prevent a crime against science by letting HST die earlier than possible. His other point strikes closer to home with me, when he says that if NASA cannot convince itself that a shuttle mission to the Hubble safe, when such a mission has been done five times before (once to deploy it and 4 times to repair the observatory), then NASA is fooling itself if it believes that it will send people to the Moon or Mars. While a trip to the moon has happened before, the second mission will have risks that we don't even understand yet, and will provide a much greater challenge with a safety bureaucracy in place.

On the road back, NASA has added a flight to the manifest, STS-114, which will not be a dedicated space station mission, though a stop has not been ruled out. I believe that the shuttle will stop at the station on this mission. The primary focus of the flight is to try all the fixes that have been implemented. The full flight plan is still developing, but it will likely involve several EVAs to demonstrate a new access to the bottom of the orbiter, as well as a number of repair demonstrations being done to multiple types of thermal protection system materials within the shuttle cargo bay. I'm not aware of any reason why this flight couldn't spend the majority of its time in the vicinity of the space station, as it would provide excellent photo opportunities to show people back on Earth the fixes in work. Before or after all the tests were run, the shuttle could drop by the station to leave off some likely badly needed equipment and supplies.

NASA's efforts to return the shuttle to flight have cost more than originally expected, and taken longer than projected at the onset. In this case, NASA's also hit on some hard luck, with four hurricanes sweeping through Florida within a matter of weeks. Two caused direct damage to processing facilities at Kennedy Space Center, while one came very close to the external tank processing facility in New Orleans. These events, plus the daily grind of working difficult problems, have pushed the launch date for STS-114 out to at least May of 2005. This is a good sign that they're not falling into the self-set trap of marching to a timeline (US Core Complete by February 19, 2004, for example) no matter how little sense the timeline made. This break between launches after a disaster is now approaching the same length as the break between flights after *Challenger*, and some take this sign, combined with the Hubble decision, as a sign that NASA may be paralyzed by its own safety culture right now[41]. It is unclear how long that approach will be able to last, especially as the ISS's ability to support a crew comes into question.

Of course, the best answer is going to come down somewhere in the middle. An overly safe organization is not the answer, because the way to be completely safe is not to fly, but the unintentional introduction of unsafe practices is what led to February 1st, 2003. So we need a safety culture that still allows things to be done. The important thing is whether or not they'll hold on to the mindset in the months and years after return-to-flight, as NASA again reaches beyond Earth orbit.

Chapter Four

Space and Popular Culture

As much as space plays a role in the lives of those of us who work with it each day and follow its activities closely, most people in the world do not pay attention to space activities. One day, at a rather famous museum, I was talking to someone about the Mars rovers. He asked me when they were launched. When I told him that they started their journey in mid-2003, he was floored…he had no idea! Big events, especially bad ones, can catch the public's attention for short periods of time, after which the daily deluge of news in other areas drowns out the space event. Of course, attention paid to something in mass media does not necessarily relate to its actual impact on our daily lives, although the opposite could be true.

Comparing Space to "Everyday" Activity

"It's 4th down and 8 yards to go with 15 seconds on the clock. Martinson takes the snap, fades back…no one's open. He spots a hole and races up the field; dodges one tackle, now he's running up the sideline. Smith has one last chance to catch him but he misses…Martinson *scores!*"

At work yesterday, two co-workers got talking about the weekend football games. They discussed the close calls that led to one team's win, and the come-from-behind action that brought about another's victory. Next, they jumped online to check how their fantasy league standings changed because of the weekend events. In the middle of this discussion, it occurred to me that no such discussion takes place about space travel in any of its forms. Space travel is a background activity, except in the cases of an amazing new mission sending back data (think *Mars Pathfinder*) or a disaster (*Challenger/Columbia*). Let's examine some reasons why.

Space missions are pretty dry events, unless something goes wrong. Once all the smoke and noise of liftoff are over, there isn't much that's newsworthy going on. A spacecraft in orbit can usually be compared to a science laboratory or a computer repair shop, depending on the mission. Neither of those activities (active research or computer repair) have found much of a niche in today's broadcasting environment, so there's no reason to assume that people carrying out such activities in orbit would be very interesting to the viewing public at large. One of the factors leading up to this is the fact that the astronauts are, for the most part, simply carrying out the activities that they've trained for. In some cases, this training has lasted years. They're not involved in moment-to-moment decision making. When something does go wrong, in many cases they don't have enough information to fix it themselves, so they have to rely on ground teams that have the additional information to help them solve it.

There is an exception, when all this planning goes out the window: when something goes seriously wrong. *Apollo* 13 provided the best example of astronauts in peril with a happy ending, and it held people's attention for days straight. Even then, there was no live footage from the capsule, as their communications had been reduced to a whisper of audio. The other major space disasters (*Columbia, Challenger, Soyuz* 1, and Salyut 1) all happened either out of radio contact, or happened so fast that there was no action that the crew could take to save themselves. Coverage of the disaster, cleanup, and press conferences following the events have held people's interest, but not to the extent that *Apollo* 13 did.

In the future, as space flights become longer because of their destinations, the astronauts will be spending a portion of their time simply traveling. The timelines here are pretty impressive, too. While some asteroid missions have been advertised as 30-day one-way trips, the journey out to Mars will take 200 days. During the trip out to Mars there will be no science for a crew to do, and maintenance should be minimized because they're in a brand new craft that has to support them for a total of two and a half years. Here, the best news coverage comparison that can be drawn is that of broadcasting news from the passenger cabin of an airliner crossing the Atlantic. Now there's some exciting television viewing! Granted, the first Mars mission could have some novelty attached to it, especially if the crew is living in partial gravity, adjusting to Mars levels on their way out, though a proper testing program would have had another crew demonstrate reduced gravity in an earlier flight to Earth orbit. This long period of travel time will have some great public relations uses, as the crew can watch and review movies, answer letters from kids in schools, and check out new books. Let's contrast this to mass-media events.

Sports championships and award shows are major media events by most measures. Lots of people from around the world watch them, and synergistically, many other news outlets report on some portion of the event on their own time. In the case of the Superbowl, this has grown into a traditional all-day party across the United States, and nearly a de-facto day off for the Monday after, as people recover from the previous night's celebrations. Continuing to use the Superbowl as an example, this is where synergy comes into play. The huge popularity of the event itself causes advertisers to flock to it, showcasing their latest advertisements. In fact, so much is made of Superbowl advertising that the rates people pay to put their commercials on the program are news items themselves, and there are people who watch the event more interested in the commercials than the actual game.

A sporting event is relatively short, and therefore is more likely to hold people's attention. It's true that there are series games, but each installment is only a few hours each, and when a sporting event goes longer than average, that fact itself makes the news sometimes. The following of sports has evolved so much that now, even the process going into building a team is monitored by fans, and therefore the media.

In professional team sports, new members are brought on board through a draft process. In recent years, this draft has drawn increasing news coverage. I'm sure that sports channels state that they cover the draft because more people are interested in the proceedings than in the past, but I believe that some of the blame has to go on the fact that there are more sports channels today than there were even five years ago. With all the sports channels looking to fill their 24 hours of broadcasting time, it is better to expand coverage of already popular sports (football and basketball, for example) than to experiment with broadcasting less popular sports.

What's the draft process like in space travel? Unfortunately, there's no set criteria that says the best astronaut gets chosen for a particular mission. In fact, some accounts report that flight decisions are based much more on interoffice politics than anything else[42]. As we'll cover in a bit, the astronaut application process is a bit nebulous, with some people applying over and over again without knowing the reasons why they've been denied. Suddenly, they may be accepted into the program even though their resume has not changed that much since the last application.

What if that changed? What if an astronaut class records were posted online? Test scores, simulation results, peer reviews, all there. Granted, this would only hold a select group of people's interest, but it's possible that this transparency would allow the number to grow over time. As important missions came close, discussions about which team would do the best job would ensue, raising interest levels even more. Of course, the ultimate read on the popularity of an activity relates to whether or not you can bet on it where such things are legal. Anyone with any allegiance to space travel as it's done today will scoff at the idea, but imagine placing odds on a new astronaut class as to whom would be the first person to land on an asteroid!

A scenario like this could be tested through a television show that combines today's reality TV fascination with the power of the internet. A group of real people could start the program, going through training for various roles in a fantasy mission to Mars (always my default goal). The first season would focus on their training, along with the personal interactions between the candidates. The season ends with a press conference, in preparation to announce who will fly on the first mission to Mars. The announcement isn't made, however, because fans choose the crew during the break between seasons. The following season picks up with the crew training for their flight as a team, and dealing with problems with their systems, publicity appearances, and other teams' desire to jump in and take their place, all the while holding together family life at home. Their launch to Mars would likely take place in the second season, followed by a landing in the third. If interest remains in the program, the previous crews that weren't selected for the first mission would continue to train for the following, and the entire process could repeat.

There's another, likely simpler method to experiment with this kind of change in space coverage. Currently, in the backwaters of most cable systems, there's a station called NASA Select. I've received the channel before, but usually only found it as part of some channel-surfing odyssey of mine on a boring Saturday. The station has live footage during space shuttle missions, usually showing a split view of mission control and computer-generated images of the shuttle and its goal for the mission, such as Hubble or the ISS. The station also goes live reporting other space events such as NASA related launches and tests. When a major (or even minor) press conference takes place, it's usually covered by NASA Select. When major events aren't going on, NASA Select reverts to playing a stream of historical space footage or some educational videos about basic spaceflight principles.

While NASA Select is interesting at times, its overall production level is rather poor. There are no newscasters, with shuttle and ISS narration relying on the voice-over of a public affairs officer, and any fancy graphics generated as segues between events are overused. It's likely that overriding policy is that anything more than this CSPAN-type coverage would cheapen the activities being covered. It's also likely that the money set aside for NASA Select is so small that larger production budgets are not possible, and, of course, since NASA is a government agency, they wouldn't be allowed to sell advertising for their broadcast. So, what are the alternatives?

There are interesting things going on at NASA, even when there aren't humans in space. At various centers across the country, new space projects are in different stages of work, with elementary research and development going on as well. Contractors around the nation are building and testing spacecraft, and future missions are being considered while policy is set for long-term strategy. At the same time, there are dozens of astronauts in some phase of training for their upcoming missions in flight simulators, giant swimming pools simulating zero gravity, or in a space station simulator. Assuming that the current goal of maintaining human presence in space holds up, there will always be a mission ongoing, suitable for at least a fifteen-minute wrap-up at the end of the day. A few well-placed cameras in a couple facilities could provide hours of interesting footage with voice-over commentary and additional illustrative graphics giving the whole thing a polished look. While the viewership won't hold the numbers of a major cable company's newly produced movie, it should be able to command more interest than some of the other specialty cable channels out there today. If the footage were provided in a constant feed, NASA wouldn't even need to run NASA Select anymore, and another entity could take it over and make a profit on its advertising.

This kind of insight into the everyday lives of astronauts, engineers, and scientists would provide a much more complete picture of life in the space program. Some would argue that this would actually serve as a disincentive, but I disagree. Yes, there will be some dull times, but when those dull times hit additional information can be added by interviewing astronauts and experts about what's happening, the same as is done today in regular news coverage.

There was one attempt to do this type of coverage of a space mission that I've seen. In 1993 the Space Shuttle *Endeavour* launched on the first Hubble repair mission. I remember watching CNN as it tried to make coverage of the event exciting. I give them credit for trying, but without more preparation as to what the astronauts were doing at any one time, and what types of problems they might be facing, the news "event" turned into a series of phone interviews with people who had increasingly tenuous ties to the operations going on. One particular low point came when the anchor covering the repair asked a VP at the company that produced the space suits (I could just picture this person being woken up for the interview, sitting on the side of his bed) about what would happen if the suits were to get a hole in them.

Unfortunately, this event is out there and it's likely to be brought up whenever some serious effort at real-time space coverage is pitched. I think it will be used as a reason why such coverage wouldn't work, instead of taking an objective look at it to figure out how it could be improved to hold the interest of a small audience. As time goes on, and such coverage becomes more commonplace, the audience may grow.

Differences Between Space Activities and Sports

Sports participants have obvious skills and talents that they are paid to show off for people to watch. These talents, through time, have taken on a higher stature in life if news coverage is a gauge. Space travelers are chosen for their talents, but then their talents are underutilized. For example, if someone has a PhD in biology and they're sent on a space shuttle mission, their skills are underutilized because the majority of the work is done by others on the ground preparing for and interpreting the results of the experiment. In space, the PhD largely serves the role as a technician, turning an experiment on and perhaps taking measurements as the experiment proceeds. While the lure is that the astronaut needs a PhD in case something goes wrong and the experiment can be corrected, this is unlikely (not impossible) because once the initial conditions of the test are changed, the results will be suspect. This is only one difference between being an astronaut and being a sports star.

Many people watch sports that they participated in during some part of their lives. Be it pee-wee football or little league, many players who had dreams of becoming sports stars now participate virtually, watching others play the games and commenting on what could have been done better. This is one place where the space program is sorely lacking. It's true that there are some space camps around the country, where children can participate in simulated space missions more or less affiliated with NASA in one way or another. Since the *Challenger* tragedy, a non-profit organization came about to expose children to space travel ideas on a smaller scale. Called The Challenger Center, it gives children a chance to go on simulated missions to Mars, the moon, or a comet. While these activities are definitely steps in the right direction, their numbers are dwarfed in numbers and overall depth by youth sports participation.

Sporting events produce unknown outcomes of little consequence. I realize that the last three words of this subsection will rankle some hairs, but on the grand scheme of things, a team winning a sports match, even for a national or world championship, has (maybe "should have" is a better choice of words) little or no bearing on our everyday lives. Yes, anyone who made a wager in the winning team's favor will be richer, and yes the water-cooler talk at work will be very animated describing the reasons why the champions defeated their lesser opponents, but life doesn't change. Here's a real zinger: in its current form, the same argument can be applied to space travel, although I feel you need to include the term "short term" in my initial argument to make it apply to space travel. Space travel plays a very small role in the general public's short-term day-to-day lives, but I believe that it has tremendous promise in playing a big role in the future. Sports, on the other hand, likely holds the same importance today that it will hold for quite some time.

The important thing here is that when the kickoff, first pitch, tip off, or puck drop takes place, no one can state with certainty what will happen. While absolute certainty is not the case in space travel either, it's much more likely that a space mission will complete to some sort of satisfactory conclusion than a team that's chosen to beat its opponent by some number of points will do so. Upsets are the reason many people watch sports, and space missions with unexpected complications are the equivalent of upsets. Unfortunately, recent space upsets have been decided in seconds, with no chance for the crew to pull a diving catch and save the game once their problems were noticed.

People can identify with sports stadiums, and the basic layout of a field, so the place where the sports contest happens is easy to visualize. Cramped quarters of spacecraft where things that people take for granted like gravity don't have their usual visible affect may be a bit unsettling to people. This could be turned to an advantage, when large structures are placed in space and sporting events (either entirely new ones or those adapted from their Earth counterparts) take place. This will cause a whole new round of discussions of new strategies and bad calls.

Because sports figures are seen doing things that constantly display their skills, it's easy for people to picture the amount of work that went into them. For an astronaut translation, about the only thing that fits well in this form would be piloting skills, since most people have some feeling as to how something flies. Unfortunately, once again the fact that things rarely go wrong enough in space to be serious steps in again. A pilot's ultimate challenge is the landing of a space shuttle, yet an orbiter coming in for a normal landing is a pretty low-intensity activity, even though at any moment if something went wrong it would become very high-intensity. The excitement of a landing is hampered by the fact that coverage is restricted to the view from the outside, and even if people could view the inside they'd only see a group of people sitting in chairs, occasionally pressing buttons or moving other controls.

As mentioned before, sports require just a few hours of someone's time. Even if a fan misses a particular event, they can take a look at a box score in a newspaper and gather the pertinent highlights of a game, at least enough to talk about it with his buddies at work. Space missions, especially those of long duration, are tough to summarize in a couple bullets. There are definitely times of added excitement, though, such as liftoff, landing, and first trips outside, though in the first two cases computers are doing almost all the work and the travelers are mostly on board for the ride. Outside of those activities, don't expect much action.

If someone gets injured on the field of play, a group of medics come out to help him or her off the field. During any break in the action, other participants take on water or discuss strategy with coaches or each other. In football games especially, the field is lined with photographers and videographers, capturing the event for all manner of later reporting. Coaches can be seen talking to their assistants standing at a distance, gaining a perspective that they're perhaps missing from the field level. These are all examples of support personnel who are very obvious in sporting events. In space missions, the support personnel are all on the ground and their only connection to the home team is a tenuous radio link. The mission control people, featured in movies such as *Apollo 13*, are just the very front lines of an army of support working in other buildings and in many cases around the country as a mission unfolds. They fit into the background very well, which is useful for letting them get their job done, but their news coverage is very limited.

So, there are some parallels between space activity and sports, but sports catches much more interest in popular culture. While I don't believe that space missions can build up the type of following that a championship game develops, I think that there's plenty of room to grow in the area of space coverage.

MODERN-DAY MYTHS

The ancient Greeks, Romans, and other civilizations had their mythologies, describing the creation of the world and humankind's place in it. Today, we have our own set of myths related to the space age. Some of them are out and out false, while others contain hints of truth spread throughout misconceptions. Others are only true right now because of the way we choose to exploit space, and the concepts would change radically if we approached space in a manner more like things we do every day. Let's take a look at some of these myths.

The Astronaut Myth

> "Astronauts work all their lives to be in space, so the idea of rich tourists paying their way to fly along with them is wrong."

One day, soon after the *Columbia* breakup, I was talking with a co-worker. Our discussion turned towards astronauts and how many we'd lost, and she immediately jumped to a statement that took me aback. She said, essentially, "All I know is that teacher (Christa McAuliffe) had no business being on *Challenger* when it exploded." I asked her why, and her answer related to the opening quote of this section. She also said that spaceflight was too dangerous for normal people to fly. I asked her when she felt that normal people could fly, and she didn't have an answer. To me, this is a problem. If we've elevated astronauts to some pedestal so that only they can make the journeys into space, we're doing the entire space effort a disservice.

I have to tread lightly here. I've met several astronauts in my travels, including both the "old school" guys (Buzz Aldrin, Harrison Schmidt, and John Young) and the "modern" astronauts (Eileen Collins, Franklin Chang-Diaz, John Casper, and I'm pretty sure I had a passing encounter with Michael Anderson, who died on the *Columbia*. The fact that I'm not sure is an interesting testament in itself) and I think they're great people. I've applied to be an astronaut before, through the Air Force, who requires all active duty officers to be ranked according to their system before being sent on to NASA. Given the new space initiative at NASA, I intend to apply again when applications are accepted. Needless to say, in my previous application I didn't make the cut because my home address is not Houston, Texas.

Astronauts are not, and I believe that all of them will agree with me on this, superhuman. When an astronaut gives a presentation, many of them go out of their way to point out how ordinary their lives were growing up, yet the "astronaut as a superhero" and "only they could do the things they do" myths live on. Let's take a look at what distinguishes an astronaut from the rest of the population.

Education

I'm going to go out on a limb here and say that astronauts are well educated. When I applied to be an astronaut while in the Air Force, I met the minimum educational requirements by having a master's degree. Many astronauts have PhDs or multiple masters' degrees, and some have more than one PhD, or are doctors that also went into another field of study. So, the astronaut corps is more educated, on average, than the rest of the population.

Does this put them on a pedestal? Perhaps, but the pedestal is not very high if so. According to the 2000 US census[43], 5.8% of Americans have masters' degrees and just under 1% have PhDs. Since census data does not divide degrees by type, we'll assume that 10% of those advanced degrees are in science or engineering fields. This creates a potential pool of 1,250,000 astronauts, yet only 103 people are serving as astronauts right now. So there has to be another distinguishing characteristic.

Piloting Skill

This has become the hallmark of the astronaut. In the beginning years of the space program, I believe that it was a necessary skill for the job. Pilots, especially test pilots, knew how to deal with high technology equipment, and could adapt quickly to changing situations that everyday people couldn't imagine. In the days before the Mercury flights, where Americans first flew in space, there were only guesses as to what kind of conditions the astronaut would face on their journey. Also, the machines were new. Mercury capsules flew a total of 6 times with people on board, and used a brand new capsule each time. Aircraft receive thousands of hours of test time before going into production, and the first flight of any new aircraft is considered a test. Does that mean that we need test pilots today?

A shuttle crew is typically made up of 7 people. Their roles include the commander, pilot, mission specialist, and payload specialist. The addition of Russian crews and tours on board the ISS have added terms like "flight engineer" and "Soyuz commander" to the lexicon, but for American astronauts, only two of the crew need to be pilots, and they sit in the front seats of the orbiter during liftoff and landing. One of the two actually carries the title of pilot, while the other holds the more prestigious moniker commander. There are probably some subtleties here that aren't apparent to the public, such as crew rotations being faster for pilots and commanders than mission and payload specialists, but given reports of the selection criteria for missions, in some cases at least there's little logic involved beyond which astronaut hasn't upset the brass at Johnson Spaceflight Center.

The other five positions do not require a pilot's skill, yet a significant percentage of those astronauts are pilots. The argument used is probably that the space environment remains unpredictable, and test pilots still have the best repertoire of skills to deal with those unknowns, but I think it has more to do with the fact that the first groups of astronauts chosen were pilots. When the time came to choose more astronauts, it was hard to argue with success, so more pilots were chosen. Over time, this focus has decreased, but I'm not sure that it will ever drop to the levels required by missions (2 out of 7 astronauts being pilots).

Physical Fitness

Astronauts' physical fitness levels are legendary. While providing some comic relief, the medical scenes in *The Right Stuff* only touched on the medical examinations done on astronauts in the early days. Again, this was necessary when people didn't know what conditions a space traveler would face when they arrived. Other considerations included the fact that missions were very short, and an astronaut with a cold would not operate at peak efficiency. These were valid concerns in missions lasting hours or days with a very full schedule of activities choreographed down to the minute, maximizing events on the trip.

Now, missions are changing. It's true that there are still missions on the books where the crew splits into two shifts and works a 24-hour cycle to conduct experiments (*Columbia* was on such a trip in 2003) but those missions are becoming rare. The majority of time spent in space now is on board the ISS, where tours are measured in months. On that type of timeline, there are days of decreased work (days completely "off" are unlikely, as there will always be some housekeeping tasks required). Missions beyond low Earth orbit will have downtimes during travel, ranging from a few days for lunar missions to over 6 months for Mars trips. So, overall health will be a consideration, but knife-edge, absolute perfect health will be less so in the future.

Even the physical demands of missions in the '60s and '70s were different. Alan Sheppard, America's first flying astronaut, was exposed to 11 times Earth's gravity (known as 11 Gs) during his mission. That is an amazing physical feat that few people could, or would, care to duplicate. Today's space shuttle exposes its passengers to a maximum G force of 3. It's true that the Russian *Soyuz* capsule exposes crew members to higher G forces, and it's likely that the upcoming, capsule-based spacecraft built in the United States will do so as well, but they will not be as extreme as in the past.

Work Experience

Outside of piloting skill, work experience for an astronaut is hard to quantify. Physical science work is definitely a plus, but it doesn't appear to be necessary. One "back door" method to becoming an astronaut is to be a principal investigator on an experiment that will fly aboard the shuttle. In that case, you'll typically only fly once when your experiment is featured, but hey, who'd complain about that? Examples of such astronauts can be found on the NASA website[44].

Another method to becoming an astronaut that's become available lately is the educator astronaut. Announced in January of 2003, nearly 1,300 teachers were nominated in the first week. By June, 1600 people had thrown their names into the hat. A press release[45] states that there will be three to six educator astronaut slots in the first class. This program is different than the Teacher in Space program that selected Christa McAuliffe, because the teachers chosen will be trained as full astronauts and deal with educational concerns on the side. Ms. McAuliffe was brought in primarily as a teacher for her mission.

Military Experience

Once again, the high percentage of pilots on the astronaut roster skews the numbers a little bit here, but there are a lot of active and former military personnel in the astronaut corps. The argument probably goes that military people are used to working in a structured environment, and are able to follow orders well. Having been in the military, I've seen that be the case most of the time, but I think another player is the same thing that keeps pilots at such a high percentage. Many astronauts in the past were in the military, so the bureaucratic inertia is geared towards keeping it that way.

One ironic thing about applying to be an astronaut from the military is that the military has its own selection criteria that a person must meet before their application is sent on to the NASA selection board. You might think that this is good, with military boards weeding out candidates who wouldn't have a chance against the NASA criteria, but the military doesn't always have the exact idea of what NASA is looking for with each call for astronauts, so sometimes this cuts the number of astronauts selected from a particular service. As mentioned before, I applied for just such a slot in 1999, and did not make the cut at the Air Force board. The funny thing is that when the list came out, a Thomas D. Hill Jr. was listed, and people in my office were convinced that I was in. Unfortunately, my middle initial is R, and I'm not a junior. If the list came out with one of these differences, I would have called to verify that they hadn't gotten my name wrong.

Desire to be an Astronaut

It's entirely possible for a person to work an entire career in academia, making outstanding contributions in the fields of physics, astronomy, or chemistry, and never desiring to be an astronaut. Someone can have an outstanding military career, flying every type of aircraft in the service, without wanting to fly the shuttle. There are those, however, whose only goal in life has been to be an astronaut, where all the actions they took in their life had that singular focus. I've exchanged e-mail with one person like this, and while I envy his drive and focus, I hope that continued non-selection, if it comes, will not be too much of a disappointment for him. On the flip side, I also hope that he is not disillusioned with the astronaut career if he's ever selected.

Patience/Persistence

Here's the one area astronaut candidates have under their control that can make the difference. As I write this book, I'm aware of two people who've applied to be astronauts repeatedly. I served in the Air Force with one of them. Denette Sleeth is her name, and she applied every time the opportunity came up, figuring, "One day, they'll get tired of seeing my name and pick me." Another friend of mine, Al Muscella, recently gave up in his application process, though he plans to reapply under the new exploration program at NASA. He had an active application open for 5 years. Recently, however, in an effort to cut the astronaut force, NASA stopped selecting for new astronaut training classes. In what I consider to be a bad move, even while it had to be done, NASA also took their last graduating class of astronauts and changed their status to "astronaut candidates." Even though there's been no board for the last couple years, Al gets a letter from NASA every 6 months, asking him for a complete update to his application package. Putting one of them together is a time-consuming process, even though it gets much easier once you've done it the first time. Eventually, the diminished chances of meeting a board along with the repeated requests for updates pushed Al beyond his limit. He ignored his last request for an update, and assumes that he's completely out of any running now.

Sad stories, but if NASA wanted you, they'd take you, right? Well, there are probably other factors at work.

Luck of the Draw

Here's the big player, that indefinable quality. Three thousand people apply to be astronauts every time there's a board, and on average, ten people are picked[46]. Of course, it's been said, "The harder I work, the luckier I get," and this will hold true in astronaut applications, too. Anything that makes you stand out above the other people who are applying will be a plus, and the only way you add to your resume is by trying new things. Who can say, though, what will catch a board member's eye and make you the candidate that they're looking for that year?

Political Power

There are some astronauts who've flown exhibiting little, if any, of the above. It wasn't widely publicized, but two congressmen, active at the time, flew on space shuttle missions. Senator Jake Garn[47] was along for the ride when *Discovery* flew in April 1985, and Representative (now Senator) Bill Nelson[48] flew on a mission that same month. Both were given the title "Payload Specialist," though their online bios don't mention which payloads they participated in.

You'll note that I left John Glenn, who flew on the space shuttle in 1998, out of the previous paragraph, and I did so on purpose. In my opinion, John Glenn deserved to fly on his second flight to space in 1998. My reasons don't have to do with the scientific objectives described as the justification for his flight, but with the fact that he was essentially grounded as an astronaut after making his first journey in 1962. According to news sources at the time of his second flight, President Kennedy told NASA that John Glenn was not to fly again. The rationale was that the loss of such an American hero would put Kennedy's vision of landing a man on the moon in jeopardy. I'm not sure that such a loss would have put that goal in danger, given the other obstacles that it faced along the way (including, unfortunately, the death of the man with the vision himself), but the decision was made. Personally, I'm glad I got to witness a spaceflight with John Glenn aboard.

Political favoritism doesn't only apply to those leading the United States. Sultan Salman Abdulaziz Al-Saud from Saudi Arabia[49] flew on one shuttle mission, though I'm not familiar enough with the Saudi political system to say that the title "Sultan" means anything special. His bio describes his leadership in several Saudi philanthropic organizations, and then mentions that he helped ARABSAT deploy their satellite during shuttle mission STS-51G that flew from June 17-24, 1985. None of his philanthropic experience involved satellites.

Astronaut Training

Once you're selected to the astronaut program, you take on the title Astronaut Candidate or Ascan. You enter a general astronaut school with others selected the same time as you, and you spend about a year learning about the space program and your role in it. Technical training involves SCUBA work, rides on the zero-gravity simulator, also known as the Vomit Comet, survival training, as well as some basic, software-based work within your specialty. Pilot astronauts differ a bit from this in that they get to fly shuttle flight simulators and shuttle flight trainers, which are aircraft modified to have the flying characteristics of the space shuttle orbiter on return to Earth. In order to simulate orbiter flight correctly, the engines are actually reversed slightly.

And What's it All For?

So you've been selected as an astronaut, and gone through the basic training. Some say you've worked your whole life for the opportunity, but hopefully I've convinced you that's not always the case...but what do you get as a reward?

Well, it turns out that as a PhD or a master's holder, you get to get in line behind dozens of other PhDs and masters holders, awaiting a flight slot. This could take anywhere from months to years, depending on your skills vs. what's required for a particular mission. Another factor is the flight rate, the number of flights in a year, and that's been decreasing over the years. It's been described that, at least in the past, very minor infractions of unwritten codes could jeopardize your position in any flight rotation. In the meantime, you volunteer for some jobs on the ground, proving that you're a team player and willing to take on the tough positions. Finally, if you're lucky, your big day comes...you've been selected for a flight!

This is where the real grind kicks in. Now you start preparing for your specific mission. This is also where a recent article in *Popular Science* magazine comes into play. In October 2003, *Popular Science* profiled the 18 worst jobs in the science field. The article was self-described as unscientific, and was very tongue-in-cheek, but held at least one revelation when it listed the job of astronaut as number 14 in the worst list. Other items on the list included Postdoc (someone who's been awarded their PhD, but doesn't hold a teaching position yet, so they're usually treated as well-educated cheap labor), hot zone superintendent (a person who repeatedly works in laboratories designed to handle the deadliest, incurable diseases), and on the poignant, tongue-in-cheek side, metric system advocate (no description needed for anyone who was in school during the '70s, but for those who weren't, there was a seemingly powerful effort then to convert the United States to a nation that used the metric system exclusively. Now, that effort is down to two part-time people at the National Institute of Standards and Technology). The article rated each job based on several categories, and astronaut achieved its ranking due to the risk of death (*Columbia, Challenger, Apollo* 1), psychological torture (repetition of the same tasks in training), physical torture (if you're on a mission where some form of medical test is being done, your physical comfort is not taken into consideration if it will get in the way of data collection), and dealing with a digestive product (since, as a rule, everyone on board has a master's degree or better, who gets to clean the toilet? Also, any animals that fly into space will need to have their cages cleaned).

I tend to describe the astronaut position as one holding a lot of prestige from the public, while being mind-numbingly boring for the most part, dangerous at times, with an unbeatable working location when you actually get to fly. That prestige level has dropped quite a bit, since the days when someone could easily memorize the names of all the astronauts in the program, but an astronaut attending a conference or giving a talk will still typically draw a crowd.

Once again, let me be clear: I do not intend to bash astronauts here. They are great people who've shown their ability to handle an amazing amount of diversity through their lives. In my opinion, though, treating them as something greater than the are—hard workers who applied, got lucky, and were chosen to join the astronaut program—would be a disservice to space travel in general. That's something I don't think even the proudest astronaut would want to do.

The "Man-Rated" Myth

The Crewed Exploration Vehicle (CEV) is in early development right now, and one of the big topics discussed about getting it flying is "man-rating" the Evolved Expendable Launch Vehicle (EELV). Man-rating essentially means that the booster people will strap into is somehow safer than the normal booster used to fly satellites into orbit. NASA has been somewhat inconsistent with its definition of man-rating, however.

Early in the space age, rockets were hard to come by. NASA, first facing the task of getting men into space, and then having their mission expanded to landing a human on the moon, had to use what was available. The first human missions into space until 1966 for the United States, and all the human missions into space that originated in Russia/USSR), used upgraded missiles for their trip. The upgrades varied. Some things that were added that made very good sense included an improved monitoring system so that astronauts got a better feel for how their vehicle was performing during the eight-or-so-minute ride into space. In case things went seriously wrong, the boosters were equipped with systems that allowed the astronauts to escape from the vehicle. In a little more esoteric sense, there was the ground quality check, which basically meant that NASA engineers supervised assembly and checkout activities on the rocket, and could then certify independently that it was ready to fly. It's important to note that this is the type of independence that suffered as cost cutting worked its way into the shuttle program in the mid-'80s and late '90s. NASA engineers, not as involved in the process as they'd been in the past, were forced to ask their contractor personnel for a certification of the spacecraft's readiness for flight. The contractor said that it was ready to fly. Big surprise. Apparently, the original approach did something right, because the overall rocket success rate in the early '60s was not stellar, but every US booster that carried astronauts performed well enough to complete their mission.

By the late 1960s, the Saturn boosters built by German ex-patriot Werner Von Braun were ready to fly. These boosters, though designed mostly from scratch, were very similar to the ballistic missiles that early US space flights started with. Since these rockets didn't have the pedigree of earlier boosters, there was some debate as to how many flights of a craft had to take place before it could be considered "man-rated." No one wanted to put people on a rocket on its first flight, but was the second one safe? What if the second uncrewed flight had problems? How many problem-free flights were required before people could travel aboard? These were the type of questions that arose in flight testing the Saturn V moon rocket in 1967-68. The first test was flawless, while the second experienced several problems. The third flight went well, and the fourth flight sent *Apollo* 8 on its Christmas voyage around the moon.

Inexperience with a booster is definitely a reason to be concerned about its safety, but I'm convinced that another factor was at work during the Saturn V certification process. In 1962, when boosters were rare and the need to get someone into space was pressing, the Atlas booster that took John Glenn into orbit was the only option. NASA was also relatively young in its organization, and hadn't gotten its bearings with full bureaucratic review of activities before they take place. As time wore on and the Apollo program came around, the bureaucracy was better formed, and had found its ability to say "No" or at least "Well, maybe...."

It's interesting to see how this argument evolved over the years to the first space shuttle flight in 1981. That mission flew with people on board. There was no mention of "man rating" (by this time, "woman rating" was also a possibility) the shuttle stack before putting people on it. Many design decisions made in development made that impossible. So, for the first time that spacecraft left the earth, a crew was on board.

Despite all the discussion of man-rating the stack, the space shuttle took a definite step back in the overall idea of safety in one area: crew escape. The Mercury and Apollo capsules carried their own "get away from an exploding rocket" ticket in the form of an escape tower poised atop each capsule. The Gemini program took more of a fighter-aircraft approach by putting ejection seats into the capsule. Luckily, none of these systems saw use in saving a crew's life in the United States, but a system similar to the Mercury/Apollo escape on board the Russian *Soyuz* vehicle has saved the lives of its crew. In 1983, a booster disintegrated below three cosmonauts headed to the space station Salyut 7. The escape tower carried them away from the fireball and allowed them to walk away from the event. The space shuttle, because of its side-by-side design, is not as easy a candidate for such an escape system. It is true that some early designs featured crew escape pods, and the idea came back again after *Challenger* and *Columbia*, but once the budget numbers come through such concepts are discarded.

One advantage of returning to a capsule-based spacecraft for exploration such as the Crew Exploration Vehicle is that launch escape will be possible again. I think that's a very good thing.

The "One-of-a-Kind" Hardware Myth

This myth is a little more complicated than the others. Right now, it is absolutely true that most hardware built for space travel is one-of-a-kind. Some notable exceptions are the Iridium communications satellites, which were built in a large enough batch to require many mass-production methods to be applied. Therefore, they are essentially the same across the board. It's likely that the Global Positioning System (GPS) satellites are similar. Even though their batch was about 1/3 the size, they were produced and launched at a relatively high rate, so there wasn't a lot of specialization. One other example of mass-produced satellites may include the Corona satellites, used for reconnaissance of the Soviet Union until 1974. These spacecraft returned film canisters to Earth, and therefore had a definite life span that was relatively short.

To varying degrees, most other hardware used in space is hand-crafted, pieced together in batches of 5 or less in facilities that make the Swiss watch works of the past look like playgrounds. In a spacecraft construction facility tour, you either stand behind glass and witness the skilled craftsmen as they meticulously assemble and test their works of art, or you put on the trade uniform—cover-all, and in some cases booties, hair cover, and the ultimate talisman: a grounding strap (to ward off the evil spirits of static electricity—a valid concern, but the imagery of calling a grounding strap a talisman was too good to pass up). Once you're in the facility, you note the homage paid to the equipment labeled as "flight hardware," and know that if you touch it with your unclean hands, it will have to be re-cleaned and certified flight worthy all over again.

In human spaceflight hardware, or even deep-space exploration systems, the production rituals are even stricter. Here, there is likely to be one craft constructed with, perhaps, another full vehicle built to test how the craft will stand up to portions of its trip.

Needless to say, an organization that builds something under such conditions over the course of years is going to be very possessive of the object once it's complete. With notable exceptions (during the writing of this book, a $250M weather satellite was dropped in its factory, for example) the care that goes into any action taken on a spacecraft borders on the paranoid. "You're going to put MY satellite on top of YOUR rocket, when YOUR rocket blew up three times in the last 100 flights!? Not until I send a team of 100 of my finest engineers to verify that you've solved those pesky problems from years ago...." Never mind the fact that very few rockets have enough flights to have a statistically significant flight record. Even rockets that have been around for decades bear very little resemblance to their forebearers. There have been many changes in performance (and therefore size), so that success rates do not directly apply to the different vehicles.

So I think I've established that there is a one-of-a-kind feeling about spacecraft, but the title of this section calls it a myth. Why, do you ask? Because the reason these conditions exist is the current approach we're taking to space flight. As discussed before, launch costs are high because we don't launch enough. Due to the high cost of launching, we build satellites and other space hardware in small batches, which make them expensive. That expense and the difficulty in getting a ride into orbit drive us to make satellites more and more reliable…and the circuit continues. This is a consistent theme brought up by Rand Simberg at his website[50].

What else is there? you may ask. It's a little hard to imagine from our perspective here, just as today's computer world would be hard to describe to someone working on one of 10 building-sized computers in existence in the 1950s. Rand paints a picture, though, and I'll fall back on his description. Right now, we have two types of weather satellites, one type is parked in geosynchronous orbit, essentially hovering over the same spot on Earth all the time. These satellites provide an excellent overview of one of the Earth's hemispheres, and are wonderful for watching the track of major storms, recently demonstrated in great detail as hurricane Isabel hit the East coast in September of 2003. These satellites also provide information in real-time about smaller storm systems, such as those that produce tornadoes in the Continental United States. The cameras required to provide this detailed observation are very difficult to produce, and the satellites that they're mounted to must have sophisticated pointing apparatus to keep the cameras pointing in the right direction or at least to know which direction they're pointing. Both of these factors make these satellites extremely expensive to produce. Even with the expense of these cameras and satellites, we're limited in image sharpness by the distance these satellites are away from Earth, on the order of 40,000 km or 23,000 miles. From that distance, it's hard to get a crisp image of anything!

The other type of weather satellite resides in what's called a polar orbit, and circles the Earth 15 times every day. They orbit so quickly because they are relatively close to the surface, on the order of 700 km. These satellites provide a better long-term picture of what's happening in the world, measuring climate change and tenths-of-degree changes in sea surface temperatures. The images that these craft produce aren't as visually appealing as the geosynchronous version, because it takes the form of stripes up and down the surface of the Earth. In some cases, data from these polar-orbiting satellites have been combined to provide a global view of the world, and the results are pretty spectacular, but not very useful for real-time weather prediction. There are usually only two of any such vehicles in orbit at any one time from each organization. With only two satellites, each spot on the Earth is only visited 4 times a day. Usually, one satellite passes an area at dusk and dawn, while the other catches an early morning/early evening. When asked, these organizations will say it's all they need, but the unspoken part of that statement is "we can't afford any more." I'm willing to bet that a large number of people in any of the organizations would like to have more data.

Let's imagine the Earth surrounded by a cloud of satellites at an altitude near that of polar satellites, perhaps a little higher up so that each satellite can see more of Earth and we need fewer of them. This cloud of satellites is constantly imaging the Earth directly below it, broadcasting its data to its brothers and sisters, who beam it down to Earth in real time. Now, we've combined the real-time forecasting capability of the geosynchronous satellites with the better resolution of the low Earth orbiters. The gain here is arguable compared to the cost, but we're basing the cost on our current view of spaceflight, where satellites cost hundreds of millions of dollars and their launches cost tens to hundreds of millions. With that amount of real-time data, at a higher resolution, weather prediction will have to improve. Of course, I have a friend who's a weather officer in the Air Force, and she told me that weather predictions wouldn't be perfect even if we had a weather sensor on every square foot on the surface of Earth, so we won't hold our expectations too high.

"But what about replacements?" someone out there is saying. "If you have a constellation of 50 satellites, and some of them go bad, you'll have to launch another to replace it." This is an absolutely true statement, and the only model we have near this requirement is the Iridium cell-phone constellation, completed but never used to its full capability. There were no less than three companies founded on the business plan that launching replacement Iridium satellites could make them money. As we've said, unfortunately, Iridium went bankrupt before we found out if such a business model would work.

So the one-of-a-kind myth is true today, with caveats that it doesn't need to be in the future.

The "It Must be Complicated to be in Space" Myth

"The space shuttle has over a million parts that all need to function for a successful mission." "A spacecraft flying from Earth to Mars is the equivalent of throwing a pitch from Los Angeles and scoring a strike in Yankee Stadium." Such quotes are the bread and butter of the space industry today, made by NASA personnel and the media that reports on it. They are true to varying degrees, in that the shuttle has a lot of moving parts, but most of them have backups. The baseball analogy would be more correct if, instead of throwing a baseball, one was throwing a sphere that could change its motion a bit based on commands sent from the launch point, guiding the sphere to the perfect strike. These quotes are likely used in a warped fashion to explain why things fail, or why things are expensive. They also serve to separate the space community from society as a whole, convincing people who watch a space event in passing that it's impossible for them to take part in such a complicated endeavor. I believe that this separation does space flight a disservice in the end[51].

First off, I need to describe how difficult it is to design, build, and test things for space travel. Space is a place of near-perfect vacuum, near-zero gravity and extreme temperature differences. The ride into space as it exists today on top of rockets involves a lot of vibration and occasional shock loads where the satellite is shaken as though struck by a hammer. All of this has to be built taking into account along with the fact that very few (statistically speaking) of these devices have been built overall, and even fewer have been built by any one company or any one team. Putting all the factors together leads to a difficult situation indeed.

At a glance, the first couple items, vacuum and zero gravity, may seem like a simplifying factor. The problem comes in that some materials act very strangely when exposed to a vacuum. Some metals will literally weld themselves together, and any joint that must move for long periods of time must be sealed very well and is extremely limited in the types of lubricants it can use. Zero-gravity causes the lubricants to clump in strange ways that aren't completely understood, and there's never been a long-term (on the order of years) study on bearing life in space that then returns to Earth for evaluation. Any data on bearings has to be inferred based on telemetry that comes down from satellites on orbit, such as the speed of the object turning, the temperature of the bearing itself, and any other indirect information that can be gained through other systems on the craft.

One thing that separates satellites from other vehicles is the environment they must endure en route to their duty station. While efforts like SpaceX's Falcon are designed to make the launch environment smoother and cause less shake and shock to the satellite it's carrying, other boosters were not designed with that goal in mind. While flying to orbit, satellites can be exposed to 6X Earth gravity for relatively long periods. This happens because as a rocket continues to burn propellants, the entire stack becomes lighter and lighter. Since the rocket is still pushing with the same thrust, acceleration increases. Vibration is another factor that a satellite is exposed to, and it comes in high frequency and low frequency. High frequency vibration is better known as sound, and there are extreme acoustic tests that spacecraft are exposed to before they're launched. In the test chamber, the satellite is set next to, essentially, the loudest stereo system that can be imagined. Then, the volume is turned way up, and the satellite is monitored to see how it responds. I've not witnessed an acoustic test myself, but I've heard that you can feel the sound more than you can hear it if you walk by the test chamber during certain portions. Low frequency vibration is more like a shaking motion, and for this a satellite is put through a vibration test. Vibrations come in both random form (where the vibration happens along different axes and at different levels) and a more consistent type (where the vibration is more like a low-frequency sound, and the satellite's motion is like a muffled guitar string).

Shock is the most powerful force that a satellite will be exposed to. Shock events on the ride to orbit take place when a stage separates, or when the payload fairing (the cover over the satellite, protecting it during its ride through the atmosphere) falls away. While film footage of these events show them gracefully falling away from the rest of the booster, they're actually cut from the remaining stages by an explosive charge. This charge travels along the length of the booster and "whips" the satellite, sometimes severely. Shock forces can be greater than 10X the normal force of gravity that we know here on Earth. While powerful, shock forces are very short lived, but they contribute to the difficulty in designing and building spacecraft.

The next complicating factor is the number of spacecraft built. Production runs for everything else that we come in contact with in our daily lives is much higher. Cars? Thousands of each model produced each year. Computers? Millions. Microwave ovens? You get the picture. These items have been built en masse for years, and likely will remain so. Because of the huge numbers produced, the rigorous testing that we've described as requirements for space vehicles are only required for the first few products, after which occasional quality spot-checks are made. Now it's true that these items we're talking about are not exposed to the stressful environment just discussed for space vehicles, but their working environment is in some ways worse…they're in the hands of the general public. No amount of design, testing, marketing survey, or focus group work will predict what stress a 17-year-old kid will exert on a car after her graduation party, but, with such a large sample size (the number of cars sold overall) the failure of a car in such a situation is not newsworthy.

Spacecraft are produced in such small quantities that even the luckiest engineer working in a plant producing their maximum amount of craft will only get to work on tens of satellites. Because sample sizes are so small, each one is practically hand built and therefore the process cannot be described as routine with normal industrial standards applied to it. Again, there are exceptions to the rule, the main one being the Iridium constellation of satellites. They were produced in numbers on the order of 100 built in a few years.

Imagine a typical consumer item, such as a VCR, that was built without mass-production methods, and only in batches of 1-5 before a new model came out, with about half or more of the people who worked on the first model moving on before the second model came along. Such a VCR would be incredibly expensive and not very trustworthy.

A separate issue that impacts spacecraft design is software. Now a part of our daily lives here on Earth, software was prevalent long before that on spacecraft. Whereas the cost of memory, processor speed, and other hardware has dropped significantly for Earthbound applications, allowing for larger (not necessarily better) software, the same equipment for spaceflight has remained expensive because it's still produced at the same production rate (or close to it) as it was decades ago. This has kept software on board spacecraft (known as flight software) relatively small. Some amazing things were done with some incredibly small amounts of software in the past. The lunar module, which took humans to the surface of the moon, had the equivalent of 4k of memory. Today, a word processing file with a single space typed in it takes up that kind of space. Software in space is slowly ramping up, in that there's more of it on board most spacecraft than there was before, and with more software comes a requirement for more software testing. Unfortunately, testing for longer software programs is much more complex than testing for their shorter counterparts because as code lines are added to a program, the number of possibilities for different conditions rises very quickly.

Now that I've painted a rather dismal picture of how difficult it is to design, build, and test hardware and software for space, you're probably wondering, "What was the name of this section again? Something about complicated space being a myth?"

My point here is that these numbers and complications are the way they are because of the cycle we're in today. High launch costs drive high production costs that work to keep launch costs high. Once some type of mass production kicks in (the most realistic levels attainable would be the production of passenger aircraft, but even there the industry capitalizes on money-saving techniques where it can and still produces quality products), the enigmatic world of spacecraft design will be more quantifiable. Another term for quantifiable is statistical significance, and statistical significance will give us a much better picture of the quality of our workmanship and software codesmanship.

The "If Space Were Important it Would Pay for Itself" Myth

This is a classic argument, so by definition, it's been around for a while. The argument essentially has two parts, in that if space found its "killer application" then it would automatically pay for itself and stop draining public funds. This argument has some truth to it, although in my opinion there are a couple mitigating factors that aren't taken into account. Another part of this myth is the point of comparison—it assumes that other transportation systems pay for themselves without government support.

How many experiments launched into space have we heard about that have the potential to open up space manufacturing? Protein crystals for new medicines, ultra-pure semiconductors (kind of a misnomer, because by design, a semiconductor has some impurities, but we'll work with it) to revolutionize the computer industry, and perfectly round ball bearings produced en masse for a new generation of machines that never wear out are just a few examples that I can think of off the top of my head. "Once we get these products demonstrated, companies will want to build orbiting factories to start producing them and we'll be in space to stay." Well, these types of products have been demonstrated for at least 20 years, and there isn't an orbiting factory out there yet. The ISS may count as one, and some of its equipment may allow small batches of the above materials to be produced, but since it's not devoted to production of any one or two things, I don't count it that way. We still hear the argument, but don't see any action, so what's wrong with it?

I believe that the root of one problem lies in the misconception that simply demonstrating the possibility of a product or action immediately makes it commercially viable. There's an important distinction between possible and viable, in that something's possible when a scientist in space (or even on

119

Earth, for that matter) cooks something up in his or her lab and writes about it. Before that technology becomes viable, a number of issues have to be met: Is there a market for this new product? Can the process be scaled up to produce things in quantity? Is the cost of setup production for this new technology so high that it will take too long to make our investments back? The last question is a particularly tough one for manufacturing in space. A company that wants to mass produce something in space has to do more than build a factory (by the way, that factory will have to meet the exacting standards of the aerospace industry, and a factory's never been built that way); they need to launch it into orbit at a cost of thousands of dollars per kilogram. If the factory is to have a crew on board, the mass multiplies because of the people, their food, their living quarters, etc. After that, they need to launch their raw materials to the factory, again at a cost of thousands of dollars per kilogram, and heaven forbid they need to send people up to their factory on a regular basis for the long or short term, because the cost of sending people into space is even higher than sending raw materials. The cost of returning the finished product from space is debatable, because you can probably simply re-use the vehicle used to deliver the raw materials, but then that vehicle will have to survive a return to Earth intact. The cost of building a craft to do so isn't trivial. So, to clear this hurdle, either a product that's amazingly expensive compared to its raw material costs requiring production in space is necessary, or the cost of building spacecraft and launching them into orbit must drop by at least one factor of ten, probably more.

The second argument is one of my favorites. In it, space should take its place in line with other forms and mediums of transportation. Air travel, cars and trucks, trains, and boats all run on their own with no government assistance, so why does space have to be paid for exclusively by the public? Hold on.

Any belief that these industries stand on their own two feet is a misconception. To varying degrees, each requires some form of government assistance, or it relies on some form of government infrastructure that's so ingrained in society, we don't even think about it.

Let's start with the one that will likely raise the most eyebrows: cars and trucks. "What?" people are saying. "Cars are a source of income for governments. I pay taxes when I buy my new car, and when I follow the will of car companies and buy a car more often (or I purchase more of a car than I need, like an SUV, but that's the topic of a whole other book) than required, for vanity purposes, I pay even more into the government till." Right, cars don't require much government subsidy, but roads, bridges, highways, and police forces do. Can anyone image how much the auto industry would howl if it was required to build its own roads? How much would the price of a car go up if the cost of maintaining and building new roads were part of a specialized tax applied only to car sales? Even car manufacturers are not free of government assistance in their primary practice. In the 1980s the Chrysler Corporation requested and received a huge federal government bailout. The cause and approval reason for this bailout had more to do with the fact that Chrysler was producing the Army's M1 tank than its car division, but in public perception at least, here was a car company that couldn't make it on its own. Now, there are tremendous economic payoffs for a government to build new roads, keep the old ones in good repair, and support the auto industry when it's necessary, just as the same positive reinforcement will exist with space travel once it's allowed to become an industry instead of largely a government pet project. For now, though, it's true that anyone who owns a car cannot cast the first stone when accusing space of sapping off the government.

Air travel! Now there's an industry that stands on its own two feet! Right. Air travel faces the same types of problems that car and truck travel face, in that the airlines (or the aircraft manufacturers) do not need to build airports. Another factor in airline operation is air traffic control. ATC is a national infrastructure. I had a friend who said that everyone should become a pilot, because everyone pays for air traffic control anyway—interesting point. If you want to found a new airline, all you have to do is find a route that is underserved, purchase (or lease) some old airplanes that no one is using right now, and find some space in an existing airport. This is not a simple task, but compare that to what's required if you want to start flying into space right now. Jumping back to airlines again, think of the other infrastructure that is in place already: public acceptance of air travel, luggage handling schemes, fuel and food delivery services to airplanes, even global positioning system (GPS) satellites, ready to provide navigation signals to anyone with the proper equipment. It's true that once an airport is operating, the authority that built the facility collects landing fees to pay for the project, but the construction itself starts as a public project, financed in part or whole by the government.

So, you've started your airline, and now you can operate, using all the infrastructure in place, and immediately become a positive cash-flow business...right? Well, there have been problems with that theory as well. The United States' air travel system started with a lot of government support, and now is undergoing a series of mergers, as airlines try to become larger and spread their overhead cost over more flights. After the September 11th attacks in 2001, the government scrambled to provide aid to the airline industry, in the forms of direct financial help, loan guarantees, and an attempt to shield the companies from legal claims against them. Not all the airlines took them, as some of the more conservative carriers had money on hand to bring them through the crisis, but the numbers are huge. Unmitigated lawsuits were estimated to cost hundreds of billions of dollars, and loss in business was on the order of tens of billions. Think about those numbers compared to the civil space program budgeted at $16B a year! I understand that airlines are an important part of our economy, and I'm not suggesting that this aid wasn't justified or necessary. I present this information because an informed debate about the government's cost for transportation in the nation requires all the facts.

Trains. Here we go. I wrote most of this book while riding the District of Columbia Metrorail system, so I know a bit about train travel from the passenger point of view. The ability of a transit system to self-support is questionable. Most commuter systems are hybrids of government and commercial activities, taking the form of a Port Authority or other organization. When you switch to larger networks such as the Amtrak system, things get worse. In September of 2002, my family needed to travel from DC to New York for a wedding. We looked for economical ways to travel, and they boiled down to the following: car, plane, train. Because we didn't relish the idea of driving in New York, and that we didn't want to wait in a long line at an airport while still-relatively new security procedures were used, we decided against the car and the plane. The deciding factor in the latter decision was our two-year-old son Scotty, who's a great kid, but keeping him occupied in a security line didn't appeal to us. There was another mitigating factor, in that Scotty loves trains and busses and most forms of public transit. In the end, we paid more for a train ride to NY than we would have for a plane flight. In doing so we took longer to travel than we would in an airplane. In a purely commercial sense, paying more to take more time in traveling is not a viable business. Granted, there are mitigating circumstances like people who won't fly and families with young children who'd rather travel by train, but it's a fact that market forces alone don't drive Amtrak.

Amtrak is probably the most federally subsidized transportation system in the US. Every couple years or so, the news is loaded with stories about how Amtrak is working on some new effort that will let it stand on its own two feet. Those news stories state the fact that the Eastern seaboard portion of Amtrak is solvent (it'd better be, if it charges more money for a trip to New York than an airline does!), and if Amtrak were a true business, it would likely "spin-off" that segment of its operations to keep some profitability. There are other routes, through the heartland of the US, that do not make money, and the loss is large enough that it erases all the gains made on the East Coast. So, Amtrack goes back to Congress, asking for more money.

Shipping appears to be able to hold itself together without periodic government handouts, but a primary reason for this is that shipping's been around for such a long time. Advances in the field such as container shipping, coupled with the explosion in international trade allow more goods to be carried more efficiently. Of course, here we're talking about cargo hauling, not the scientific exploration of the oceans or research into the human response to long-term exposure to onboard-ship conditions. A change in mindset about space, to seeing it as a medium through which goals can be achieved (super-rapid air travel, planetary backup, etc), instead of a goal in and of itself (such as the much-hyped "days on orbit" for the ISS), may be part of the answer.

I don't mean to insinuate that these comparisons are perfect. Space cannot be grouped with any form of money-making endeavor yet, with the exception of telecommunications and some of the launch business. I will cheer the day when our primary methods of riding into orbit reach the status of an Amtrak, requiring an occasional infusion of Federal money, but not having a meal ticket provided each year. Of course, even when they do reach that point, it will be much more expensive to fly from DC to NY through space than on the train, one can hope… I wanted to take some time, though, to point out that industries some may think are self-supporting don't pull their own weight all the time.

The Humans vs. Robots Myth

To me, this is a tired argument, but it's one that will be around for some time. In fact it will always be with us, even as we move out into the cosmos, which I believe humans will do eventually. Essentially, the argument goes like this: "It costs too much to send people along on a space mission, when robots can do so much more for so much less money." If I've missed a subtle nuance to the plea, I apologize, but I believe that I've hit the primary meat. Let's break this argument down a little.

"It costs too much…." This argument will come up many times in arguments against any sort of space activity, so you may as well get used to it. Let's throw out some of these big numbers. A typical uncrewed mission into outer space costs $500M or more. It's true that many are done cheaper, but when you incorporate the "big ticket" missions like Galileo, Cassini, and the future Jupiter Icy Moons Orbiter (JIMO), I think that most people will agree that $500M is a reasonable estimate. This money is spent over a period of years, varying from 5-10 (in some extreme cases, such as Galileo, the $2B was spent over close to 20 years). Depending on your accounting methods, a space shuttle mission costs between $500M and $1.5B dollars. I personally put more faith in the higher number, because it comes closer to incorporating the infrastructure (launch pads, processing facilities, and the standing army of people working on the craft) than the lower number. So, for $500M, you can either buy an entire space mission to a body in the solar system, returning all the data, or you can purchase some fraction of a space shuttle flight. Sounds like we have a winner in the uncrewed mission here.

"…to send people along on a space mission…." People are hard to take care of in space. They need food, water, and air. They need to have a place to live where the temperature is relatively constant. Also, people are a messy lot. We sweat, we urinate, we give off waste heat, and a craft that carries people aboard has to handle all these things, either temporarily (such as storing by-products on board until they can be dumped) or permanently (such as radiating waste heat into space). Even worse for scientific measurements taken outside the spacecraft…we move! Any voyager who pushes on the side of the spacecraft to move across the room gives the guidance system something to worry about. If an on-board instrument is taking a picture at that time, the image will be smudged. Uncrewed missions don't have these problems, so the argument so far goes for the robotic craft instead of the people.

"…when robots can do much more for much less money." I figured we'd wrap this all up with the last phrase. Yes, robots can go places that people can't, and do so at a much smaller size. In rocket-speak, small size translates directly into relatively small price tag. By definition, if a robot can go places that people can't (the radiation belts of Jupiter, for example), then it can do more than people can.

When I give this talk to a group of space advocates, I watch their eyes droop. The argument seems ironclad when presented that way. Those who are against human spaceflight stop there and rock back in their chair, with a satisfied smirk on their face, convinced that they've proven their point. There's more to the argument, though.

The fact is that humans and robots have different strengths, and any effort in space needs to decide what it wants to do before deciding whether it should work with an uncrewed program or one with a crew. Only that initial goal decision will clarify the humans vs. robots debate. Everything else is just bluster.

If it is the stated goal of an entity to characterize the solar system and beyond at relatively low cost, then a robotic program is the answer. Robotic spacecraft are excellently suited for initial surveys of a planet, either from orbit or from the surface. Robots also have the capability to do some sample return missions, as the Stardust mission is demonstrating right now. It will capture pieces of a comet and return them to Earth for intense analysis that wouldn't be possible on board the craft. Missions have been proposed to skim the Martian atmosphere, gathering a sample of the gasses at high altitude and possibly grabbing some of the famous red dust. The craft would then swing back to Earth, with the gasses inside. Another place where robots shine is astronomy. Breathtaking deep-sky images of nebulae and planets can best be gotten from a craft with no people on board, as we've already discussed.

What if the stated goals of a space effort are greater than that? It's true that a robot can do an initial survey of a planet, but if you want in-depth exploration, a person has to be there. A robot may be designed to search for rocks, but what if an unexpected discovery comes along, for which the robot's design isn't suited? A human, with a properly outfitted lab nearby, will be able to adjust their focus and tactics for the new information. Humans will also have the upper hand over robots in the area of mobility. It's true that it's possible to design a robot to explore a cliff face, but if that robot lands off course, the mission may be over. A crewed mission with some ropes and basic repelling equipment can do the cliff-face mission, but if the landing is off course for some reason, the mission is simply degraded, not decimated.

Here, it's consistently brought up that robots are becoming more advanced every day, and there I cannot argue. Honda's recent introduction of the Asimo[52] robot is mind-blowing, so I won't go on record to say that it's impossible for robots to achieve a near-human capability. I will go on record to say that mimicking human behavior in walking, talking, and carrying out simple functions is a far cry from building a robotic rock-climber/geologist/biologist/prospector/miner.

Moving beyond the scientific curiosity of "What is a planetary body made of?" we move into the idea of "Let's get what this body's made of to our planetary body" also known as mining. Mining expeditions are inherently human activities, although on Earth they're assisted by a lot of machines. It's safe to assume that in space, robots will assist humans in this endeavor, and I won't even guess as to what the proper balance will be.

Finally, when we have small mining settlements beyond Earth, with workers moving back and forth among them, it's natural that some people will want to stay. Now we're talking colonization. Colonization is not something that can be done by robots, unless we somehow replace ourselves with them, an action I'm personally against. Harvesting raw materials, building structures and living in them is part of the human condition, and I think that it should remain that way.

Arguments for and Against Space Travel

Like the myths about space travel just discussed, there are a whole series of arguments for and against space travel. I've found that most people are pretty set in their ways as far as opinions go, and that, as usual, the more adamant someone is about their opinion the less likely they are to discuss the topic rationally. In the interest of full disclosure, as you've probably guessed so far, I'm in favor of space travel and I don't see my opinion changing in the near future. Here, I'll attempt to articulate opinions on both sides of the argument.

Since we've already covered the robotic vs. human argument, we're only going to discuss crewed space travel in this section.

For Crewed Space Travel

There are those who believe that the truest, best way to explore space is with humans. These people see the limitations of robotic exploration as serious and insurmountable and they see humans as the solution to that problem. Citing other reasons such as inspiration of the youth and technology developed for human exploration working its way into society, called a spinoff, they maintain that the cost of human spaceflight is more than justified by the positive impact that the activity has on our society. The arguments are a little more complex than that.

Exploration and its unspoken of twin expansion, in all their romantic ideas of heroism in the face of the unknown and its mythic quality in our history and writings are often quoted in the reasons to send people into space. President Bush went as far as to state that the call of exploration was written on the human heart in his tribute to the *Columbia* astronauts in Houston. As part of this rationale, many cite the fact that exploration in space doesn't take place at the expense of any currently habitable land or volume, unlike the exploration and expansion throughout the habitable lands of the Earth.

Inspiration is another reason cited for human spaceflight. The points cited in favor of this argument include the technological boom that took place during and immediately following the Apollo program. During this time, the number of science-related graduates at all levels of education (bachelor's, master's, and PhD) doubled[53]. Some graduates went into the field of engineering and worked directly for the space program, but others found other niches. It's likely that many found other niches because of their personal interest, but a definite factor in their decision probably dealt with the drawdown of aerospace activity that's taken place since the Apollo program. Anyone who's trained in science and mathematics at a basic level has a wealth of opportunities open to them, however. These include fields such as research science and medicine. Many also cite the computer/internet boom of the 1990s as a direct offshoot of the Apollo program. Any kid three years old or older who could remember Neil Armstrong's walk on the moon was a 23-33-year-old technically educated and curious individual in the 1990s, waiting to pounce on a new idea and make money on it. That willingness and capability created the boom economy of the late '90s, and provided the increased tax revenues that allowed the US government to move from deficit spending to surplus spending for a short period of time. The boom was the longest period of sustained growth in US history, but eventually deflated.

Exploration and inspiration are my favorite arguments in favor of human space activity, and therefore I don't have a lot of arguments against them. Exploration does have some difficulties in the case where we find life on another planet, and in that case it's likely that comparisons will be drawn between human expansion to Mars for example and American westward expansion. A portion of me hears the argument, but I can't compare the near eradication of Native Americans to the decrease in livable habitat for a group of bacteria that may or may not be found on Mars.

Spinoffs, or technology that we have today that we wouldn't have if the space program hadn't come along, have long been used as an argument in favor of any form of space travel. Here, I tend to fall into the camp who believes that spinoffs had better be benefiting society when a government chooses to invest as much money in something like the space program. One problem is that many of the spinoffs that are commonly cited (Teflon, Velcro, integrated circuits, microwave ovens, etc.) didn't actually get their genesis in the space program. Many of them found their first widespread use through it, though. Other, less widely known spinoffs are likely to have a stronger claim to being based on space program research, but this is only a contributor to a case for space research. In my opinion spinoffs cannot justify the costs of an effort alone.

Against Crewed Space Travel

The arguments against crewed space travel come in many forms, with passionate people describing their own reasons with their own set of backup facts, but I believe that the core arguments against space travel boil down to three things: cost, risk, and expanding a flawed model.

Cost is the most commonly quoted reason as to why we shouldn't send people into space, and there is no doubt that doing so is expensive. Unfortunately, space efforts typically cost more than originally advertised as well. There are several reasons contributing to this, some deliberate and others dealing with the expense in doing things that have not been done before in small enough batches that lessons learned in one effort are difficult or impossible to apply to another. In the deliberate category are efforts to win a government contract. No matter what kind of talk you hear about government purchasing, one truism remains: the company that says it can build the system in question for the lowest cost will win a contract to build that system most of the time. Therefore, there's a deliberate effort on the part of corporations to make their project look inexpensive. Now, once the company has been chosen to produce the system in question, they may be banking on the fact that the government will provide additional funding at a later date. I've seen this happen on more than one occasion. Another reason that costs rise in building space vehicles is that very often the government purchasers in question change their mind in what they want to do with the system they're purchasing.

I find it interesting that few people who decry to the cost of human spaceflight do not do so in the context of other government activities. For example, when President Bush announced a new vision for NASA in January of 2004, asking the space agency to complete the space station, then set about exploration of the moon and Mars, critics immediately pounced on the cost of the program. Because the President didn't give an overall cost of the program, the critics in question grabbed old numbers and, where necessary for their arguments, increased them. By the way, the old numbers that most of these critics cited, from the 1989 90-Day Report, were submitted by a NASA that didn't really want the challenge of going to Mars when the International Space Station was still in its development phase. In that plan, old ideas that have been around since the days of Werner Von Braun and his moon rocket team were dusted off and repackaged, with lots of new projects that NASA had been working for years on at a low level thrown in for good measure. The figure was about $450B, and when critics today cite it they jack the cost up to 1 trillion dollars.

A common subset of the cost argument runs along the line of saying that we should spend the money spent on space here on Earth and solve our problems here before moving outward. I believe that this argument is flawed. First, humans have never awaited a perfect society before exploring and expanding. I see human social progress as a gradual, halting motion towards general improvement. I believe that, in the world, the percentage of people living in "deplorable" conditions has shown a general decrease over the years. This general decrease took place despite the fact that all other progress wasn't halted while the human race focused on solving the myriad of problems that cause deplorable conditions.

Make no mistake, traveling in space is expensive as we do it today, and sending humans into space is more expensive than sending robots. By simple mass alone, a spacecraft that supports people brings the people themselves, along with a living space and supplies for the whole journey. This core fact is unlikely to change. However, when compared to federal budgets of over $2T dollars, spending less than 1 percent of that budget each year on human endeavors in space, does not sound as threatening or budget-busting.

Risk is another argument against sending people into space, basically saying that if we can send robots, we should. If people don't need to go, they shouldn't. As I stated before, I believe that there are roles for both robots and humans, where robots do an initial survey of a planet and humans come afterward, using the data delivered by those robots. The idea of risk folds into another, deeper reflection on society. When people on the street are asked if space travel is worth the risk, and they say "No," this is somehow forwarded as an important statement. While everyone is entitled to an opinion, I think a better way to phrase the question is: "Is space travel too risky for you to want to participate in it?" The fact that they're being asked this on the street pretty much answers the question—they've chosen not to participate. People who traveled in space in the past, travel in space today, and will travel in the future were, are, and will be volunteers. These people have looked at the risks and decided that they are worthwhile. So the risk is not society's as a whole, although society is paying for it. If people don't want to personally participate in space exploration, that's fine, but that should be a separate issue as to whether or not the activity should take place at all.

Environmental issues are another reason cited for a decrease or stoppage of space activity. "We can't even take care of our own planet, and you're suggesting that we travel to another?" This summarizes another argument against spaceflight, especially the more advanced concepts of colonization (humans actually live on another planet) and terraforming (humans alter another planet sufficiently to allow humans to live upon it easier). As stated before, some people believe that humans need to get everything in line here on Earth before moving elsewhere. Here, I find a problem with this argument's underlying assumption that society is monolithic, that all portions of society act the same way. Yes, there are sections of people within our world today who think that destroying some portion of the environment for the relatively short-term gain of profits for a company is a good tradeoff. When presented with the idea that their activities, if altered slightly, would provide a much lower impact, they cite a cost figure that prevents them from doing so. There is another portion of society, whose voice is growing louder, speaking up for the environmental causes. It's unclear to me whether the loudness is coming from more people or simply more media savvy on the part of the existing group. Given the impact that environmental interest can have on a project, and given the limelight that any human colonization effort will take place in, environmental interests will have a say in what gets sent to a new colony. Of course, they'll have very little say in what gets done with what's sent unless they have their own emissary among the colonists. In my experience, the people who are the most passionate about their cause have an inverse interest in any other. This translates to an unlikely chance that any major environmentalist will be interested in traveling to a new colony to enforce their policies and recommendations.

Another point along this reasoning may bring the human natures involved a little closer to home. In a debate between Dr. Robert Zubrin and Dr. Robert Park, this issue came up, with Dr. Park asserting the typical stance mentioned in the previous paragraph. Dr. Zubrin's response was that, if you give a teenager a car as soon as they turn sixteen, they're likely to wreck it. If you make a teenager work to save money and allow them to purchase a car, they're more likely to take care of it and value it. If your teenager is forced to buy an old car and fix it up, their appreciation for the car will be even greater. Therefore, he continued, the reason that humankind has been careless with Earth's biosphere is that we were given it, with no work required on our part. If we have to build our own biosphere, we will then treasure it more than the one we were given.

Chapter Five

Space Activism

Roots

Space activism has been around since the general realization that rockets could take humans beyond our atmosphere. The British Interplanetary Society is widely accepted as the first formal organization, and one of the founding members, Arthur C. Clarke, has dazzled generations with his visions of what is possible. The early crowning achievement of the British Interplanetary Society was a device they built that would allow a space traveler to site a star for navigation even though his or her spacecraft was spinning. Other societies came into existence in Germany and Russia, but most space activism was muted by larger events once World War II got started.

In the years of the space race, there seemed to be little need for space societies advocating governmental involvement in space travel. Some societies existed to point out other ways to explore space, while most people were content to watch the space spectaculars as they unfolded. As the party of Apollo era turned into the hangover of shuttle development of the 1970s, made worse by the *Challenger* accident of 1986, more organizations appeared to voice their opinion on what governments should and shouldn't do related to space. Some of the societies in existence today can trace direct lineage to an earlier society, while others came into being on their own, after an author found notoriety in a book publishing.

Tom Hill

A Society for Every Taste

In today's society of tailor-made activism, someone interested in space travel can find a group of like-minded people fairly easily. What's more, a person can choose their own political agenda related to space, whether it be increased uncrewed exploration, private enterprise taking over space (leaving government to their typical roles of policing and setting policy), or advocating a specific destination for humans on their expansion outward. In my opinion, this kind of division can be good because it draws people in with different ideas, but I'm pretty sure that a lack of common goals produces more confusion among the very people that these societies are trying to reach—the public and their elected officials. Imagine being a civic group member who has only a passing interest in space from day to day. You go to your weekly meeting and get a presentation from The Artemis Society about how people need to return to the moon. Two weeks later, you go to your meeting and get a talk from The Mars Society stating that the moon is a diversion on the route to Mars, and has very little value without the development of fusion. Granted, these discussions are the point of much debate in space circles, but what impression does it leave with someone who doesn't check *Space News* every day? Let's take a look.

American Institute of Aeronautics and Astronautics

AIAA is the professional society for aerospace engineers. It grew out of the American Rocket Society that got started at the Jet Propulsion Laboratory in the 1930s. There are associate memberships for educators and others who are interested, but by and large the majority of members in AIAA are student, practicing, or retired engineers. The organization has a national headquarters, and is then divided into regions, and further divided into sections. Sections are usually based in areas of heavy aerospace activity, such as the Washington, DC metropolitan area, Baltimore, or Vandenberg Air Force Base. AIAA produces a glossy magazine each month, *Aerospace America*, and also runs a series of trade journals where members can publish their work in the field. These journals are highly specialized in areas such as Spacecraft and Rockets/Guidance, Control and Dynamics, and the like.

AIAA's focus is education of and networking between its members. There is some outreach, embodied in the educator associate memberships and some educational activities with students, but the main thrust is for present and future engineers. There is a new effort underway within AIAA to create standards for aerospace engineering practices. This is a good idea, because right now there are very few standards for things that really matter, and here we're talking pretty basic stuff. For example, there is no real standard voltage at which satellites operate. If you want to move an instrument from one satellite to another type, you'll need to do a bunch of power conversion because that new satellite likely won't have the voltage that your old one used. Imagine purchasing appliances that ran on different power supplies, all within the same country!

The Planetary Society

Co-founded by famous astrophysicist Carl Sagan, The Planetary Society is devoted to the exploration of space, with an emphasis on finding life beyond Earth and extra-terrestrial intelligence. In the past, their focus has been on uncrewed, robotic spacecraft and radio telescopes for their respective searches. Recently, however, The Planetary Society President, Dr. Wes Huntress, spoke at a US senate hearing about the future of NASA and strongly advocated a humans-to-Mars mission in his testimony.

In the last few years, The Planetary Society has moved from space advocacy into actually doing some independently funded research. This research typically has a high publicity level attached. The Mars Polar lander that crashed in 1999 had a microphone on board to record the sounds of Mars, and that microphone was sponsored, designed, and built by The Planetary Society. Current efforts include the Cosmos 1 spacecraft, which is designed to demonstrate solar sail propulsion, and was described earlier.

One strong suit of The Planetary Society is its partnerships. A particularly amazing project that they got involved in was SETI@home, a massive parallel-computing effort that asks people to download a program to their home computers, and then uses extra processing time on the machine (either while the user is working, or while the computer sits powered on but idle) to analyze signals captured from outer space. The analysis looks for a signal that stands out from the background noise of space, either in its intensity, varying level, or pulsation. The response to this project was so huge, with so many people processing signals, that the SETI@home effort turned into the largest supercomputer in the world. The sheer numbers of people gave the project much more processing power than planners expected, and allowed them to add search criteria. Originally, there wasn't a plan to search for pulses in signals, but that was added later. Currently, SETI@home is preparing for an upgrade, which will allow people to split their processor time between a number of projects of scientific interest. One other option that exists that I'm aware of is protein folding, another computationally intensive research project.

Another partnership that's been in the news recently is with the Lego Corporation and their Red Rover Goes to Mars effort. This combined campaign between TPS, Lego, and NASA is designed to popularize the Mars Rover missions sent to Mars in 2003. This project had a display at the 2002 World Space Congress, and it included a full-sized Mars Rover built out of plastic Lego bricks. NASA was unhappy with the logo of the project, which included one of the Lego people, while NASA maintained that there were no official efforts underway to send people to Mars. The space agency hasn't offered a new position since the new policy announced by President Bush.

The Planetary Society boasts the largest following of the grassroots space activism organizations, with 100,000 active members. This probably has to do with the "star power" of their famous founder. Carl Sagan was the most popular scientist of the time.

The National Space and Satellite Alliance

Announced in January of 2004, this organization combined the previously individual organizations of The National Space Society, The Space Foundation, The Satellite Industry Association, and The Washington Space Business Roundtable. Billed in the press release as a combination of grassroots and industry interest groups, their goals are threefold: Building Congressional and public support for the new White House space exploration plan, working with Congress and the Administration to implement export control laws that protect our country's national security without unnecessarily burdening industry, and educating policymakers and the public about the important role that satellite systems play in protecting our homeland security.

I've worked with The Space Foundation, volunteering at their annual conference in Colorado Springs and helping them set up educational opportunities for teachers in the area. Under the new organization, it's likely that the foundation's activities will continue, just under a different banner.

Before the merger announcement, The National Space Society (NSS) sent out a number of e-mails describing some serious financial times. They were also the only space-activist grassroots organization that had an office in

Washington, DC with a full-time staff. I don't believe that these two facts are a coincidence. The cost of maintaining such an office is high, and any wavering in funding would impact the organization's ability to maintain such a presence. NSSA states that they have offices in DC, and it's my understanding that the merger has provided enough funding to keep the old NSS headquarters open.

The Space Frontier Foundation

The Space Frontier Foundation bases its ideas on the exploration models of the past, where the government carries out initial reconnaissance of a new frontier, and then provides an initial infrastructure or support need. Following that initial need (and the market that it creates), commercial interests move in, with settlers to follow.

I've heard the president of the foundation, Rick Tumlinson, speak on multiple occasions, and his favorite way to present this is by retelling the tale of Lewis and Clark. In 1803 Thomas Jefferson commissioned Meriwether Lewis to explore the Louisiana Territory, with an eye towards finding the quickest route across the continent. Along the way, they were expected to make contact with the natives that they came across, and to give the natives the news that they had a new leader back in Washington, DC. Lewis and Clark carried out their expedition, and it was, by most measures, a resounding success. They returned to great celebrity, much like our early astronauts did in the 1960s, although ticker tape parades and press conferences hadn't been invented in the early 19th century.

This is where the similarities end, according to Rick. After that initial reconnaissance, the government of the 1800s didn't press on with grander expeditions of exploration. They didn't try to invent new wagons or new boats to try and make travel to the west cheaper or more convenient. Instead, the government started building a military presence in the west, small forts manned by contingents of soldiers whose task it was to keep the peace (there are all sorts of interpretations as to what that meant) and protect burgeoning settlements and their colonists. These forts needed supplies, so the government hired services from the private sector to deliver goods to the forts. Some of these delivery people stayed in the area and set up shops inside or outside the forts. Sometimes, these settlements that surrounded the forts in question grew into cities. Cities required more supply delivery…you get the picture.

In space travel, the Space Frontier Foundation maintains, the government took the opposite approach. After Lewis and Clark returned (in the form of our early astronauts), the government continued journeying to the West, building increasingly expensive and complex ways to travel there. Because of this, there's been no ramp-up in commercial activity (one exception to this is commercial communications satellites, which I've not heard an official policy on from the SFF. More on them later) and therefore the cost of spaceflight remains high.

So this all boils down to the idea that the government, in its various forms, should constantly explore the "far frontier," while private and corporate interests should make use of the "near frontier." At this writing, Rick Tumlinson describes the border of the far and near frontier as lunar orbit, where commercial entities should be expected to make it to lunar orbit, while government agencies should take off and land on our neighbor's surface. Of course, with just a little more experience in doing that (having done it 30 years ago doesn't really count, since most of the people who worked back then have retired) would make the lunar surface the "near frontier."

To me, the analogy is illustrative, but not completely correct. At the time of the Louisiana Purchase, discussions of manifest destiny and the expansion westward were on their way to taking over the nation. While there are those who believe that humankind's future is in space, others are not easily convinced. Also, anyone with a rifle, a few friends, and some horses had a reasonable chance of trekking across the west to make a new life for him or herself. Access to space has a rather large ravine that must be crossed, being the atmosphere, before someone can plan to stay there for any period of time.

The B612 Foundation

Although not actively recruiting members at this time, the B612 Foundation[54] is working to make relocating an asteroid a high priority in the US space program. Their goal is to have the mission take place in the next 15 years. Later, in discussions of where we can go in space, I'll go into details about why this is a good idea, but for now let's just say that it is a good idea.

The National Space Club

The National Space Club was founded on October 4, 1957 (the date of the Soviet Union's launch of Sputnik 1), with the stated goal of maintaining America's edge in space and rocket technology. The club offers both individual and corporate memberships, though the maximum number of corporations that can be members at once is 54. Most meetings are held in the Washington, DC, area, and though the club doesn't meet monthly anymore, they still do hold meetings when an interesting speaker is available, and they are well known for their Goddard Annual Dinner, and the story goes that high-ranking NASA brass were at a dinner when *Gemini* 8 had its near-catastrophic thruster failure[55]. They have had a web presence reserved for some time, but just formally activated it at http://www.spaceclub.org.

The Mars Society

I know the most about this society because I'm currently a member. The Mars Society was announced at a Space Frontier Foundation conference in 1997. Robert Zubrin, coming off the surprising (to him) success of his book, *The Case for Mars*, was looking for a way to polarize that popularity into movement towards Mars. He met with members of academia and NASA who had gotten together bi-annually in Boulder at meetings called The Case for Mars, and pressed for a more structured approach. Founded on three principles, public education, political action, and privately funded exploration, the first Mars Society conference took place in Boulder in 1998.

What appealed to me about The Mars Society was the desire to accomplish something related to Mars in short order. The first project, The Mars Arctic Research Station, was approved at the first convention, and design work began right away, in which I helped out. The station was designed to look like a habitat vehicle described in the Mars Direct mission architecture, and it would be placed at the Haughton Crater on Devon Island in Canada. A researcher named Pascal Lee at NASA's Ames research center had traveled to Devon Island several times doing research on geological formations that were very similar to formations that could be found on Mars, and Pascal was a strong advocate for this research station. Despite a difficult start including paradrop failures and an assembly crew that walked out, the station was up and running by the time the third Mars Society Convention took place.

Unfortunately, the Arctic Research Station is handicapped in its inability to support year-round operations. As it turns out, July is the only full month that people can reliably reach the Haughton crater. Because of this, The Mars Society has built another station—The Mars Desert Research Station (MDRS). Another station, destined for deployment in Europe, has been built but lacks the funding for shipping and setup. Crews rotate through the MDRS year round, doing research on crew interaction, space suit utility, and geology, using methods that would be required on Mars. This is done in a simulated environment—for example, once a simulation starts, people are not allowed to go outside without putting on simulated spacesuits. Many a researcher, who put their experiment together in a lab or in some cases their garage, has been surprised by how differently they can interact with that experiment once they're wearing a simulated spacesuit. That's the reason for analog research stations.

After the Mars Arctic Research Station, The Mars Society moved on to other areas of action. In order to test mobility at the research station sites, a design competition for a rover was announced in 2000, resulting in three teams building rovers. One team, The University of Michigan, found sponsorship on the order of a million dollars for their work. Building on this success, a fundraiser was held in Silicon Valley, courting the millionaires created through the internet boom of the late '90s. At this event, James Cameron, director of movies such as *Terminator 2* and *Titanic*, and member of The Mars Society, gave a talk about his vision of Mars.

Unfortunately for the fundraiser, the internet bubble that propelled the United States' economy to such heights in the late nineties was set to burst when the event was planned, and when invitation went out the signs of the "pop" were evident on the horizon. Attendance was low, but one attendee in particular caught the space bug. His name was Elon Musk, and he wanted to know what kind of research related to Mars could be done for on the order of $10M. Mars Society President Robert Zubrin brainstormed a Mars-gravity demonstration mission where mice would be exposed to Mars-level gravity for long enough to breed one generation. Such research has never been done, with NASA focusing most of its energy on counteracting the debilitating effects of zero-gravity instead of finding ways to create gravity for its crews. The mice would then return to Earth and be examined to determine the effect reduced gravity had on mammals. Elon liked the project, and looked into carrying it out. Eventually, differences in approach caused Elon to strike out on his own, and he soon founded the rocket company SpaceX. The Mars Society pressed on with its project, and eventually The Mars Gravity Biosatellite project was born[56].

The Mars Society has had its difficulties, however. In my opinion, the organization focuses on its pillar of private exploration, relying on that effort to draw in public education and political awareness. Surely this is one approach, but there are members of The Mars Society who are interested in laying groundwork in the other two areas along the way, and that can boost the punch when an opportunity in the political or educational arenas. Towards the end of 2003, The Mars Society has changed its approach in some ways, funding a small virtual office in Washington and getting much more involved in congressional activity. I believe this will bear fruit as the nation's space policy evolves over time.

One way to check the priority of an organization is to look at the leadership makeup, and The Mars Society's steering committee is heavy in engineers, authors, and academics, but light in politically savvy people. This may change over time, as members of the steering committee are now elected by the general membership. There have even been problems within the leadership itself. For example, in a mass resignation, 4 members of The Mars Society Board[57] resigned after the 2001 convention, and some of them went off to form The Mars Institute, discussed below. There have been conflicting reports as to whether this split was a simple difference of opinion, common among space advocates, or a professional conflict.

The Moon Society

The Moon Society was founded in 2000, likely as an answer to the founding of The Mars Society. It is the goal of The Moon Society to have humans return to the moon for good, and use its resources for various purposes in the inner solar system. The last updates to The Moon Society web page[58], when accessed on February 16, 2004, were dated in March of 2003, so it's unclear how active the group is.

The Artemis Society

The Artemis Society holds a similar goal to The Moon Society, return to the moon to stay and use its resources, but they differ in approaches. The Artemis Society is focused on low-cost hardware that makes such missions possible in the very near future as a commercial venture.

In a rare (by my estimation) demonstration of solidarity, The Artemis Society and The Moon Society merged. Their websites are separate, but when the "News" link is clicked on the Artemis Society website[59], the user is taken to the Moon Society's website.

The Mars Institute

The Mars Institute officially announced its existence at the World Space Congress in Houston in 2002. Made up largely of Mars Society members disenchanted with that organization for one reason or another, the Mars Institute took what I consider to be a more academic, more formal approach to its existence from the start.

Mars Institute activities are related closely to the Haughton Mars Project. This effort started as a NASA program investigating Mars geological research through research at the Haughton Crater on Devon Island. It was through this project that Pascal Lee brought forth the idea to The Mars Society about an analog station as the society's first project, and The Haughton Mars Project continues today.

An early success for the Mars Institute was the delivery of an analog Mars rover to the research station in the Haughton Crater at Devon Island. This rover effort is entirely separate from The Mars Society's Arctic Research Station, and involved a potentially hazardous traverse across an ice sheet between Resolute Bay (the nearest settlement that qualifies as a town, largely because they have an air strip) and the crater site.

The Institute does other research in the Arctic during the summer months. They set up a greenhouse there, and have remote cameras monitoring the plants when no one is present. While the growing of plants in extreme environments will always be of interest in space travel, the endless days of the summer and endless nights of the winter remind me more of the growing conditions on the moon than on Mars.

The creation of The Mars Institute was the first split of a space organization that I witnessed, being a member of The Mars Society at the time. At first, I felt the division hurt any chances of moving in the direction of Mars. I've mellowed on that stance a bit, since some of the things that The Mars Society does give it a well-deserved reputation as being a maverick organization. This is not bad, but will not attract people who are more interested in a scholarly approach to things. The Mars Institute fills that niche. As a friend of mine once said about the split, once things get really serious, the space organizations will be sorted out through Darwinian selection.

ProSpace

ProSpace is the congressional lobbying arm of The Space Frontier Foundation. The primary event for ProSpace is its annual March Storm, where volunteers from around the country come to Washington, DC to meet with congressional representatives. In the lead-up to March Storm, the leadership and past participants work to put together a platform for the year, which is approved by vote.

Starting on a Sunday, the volunteers are trained in lobbying etiquette and the technical details of the platform. The volunteers receive schedules that have them meeting with their congressperson's staff, and starting Monday morning a group of professionally dressed people descend on the House and Senate office buildings to spread "the people's space agenda."

I participated in a March Storm in the year 2001. It was eye-opening, and if you get the opportunity to, I recommend the experience. All the congressional staffers I dealt with politely listened to our talk, but you knew that you were on to something when they actively asked questions and made counter proposals. In my case, the best meeting I had was with a staff member from the Senate science committee, who actually held us long to gather more information.

To me, the greatest revelations came as I walked around the buildings. Riding the small trains that connect the office buildings and The Capitol was one thing that was unique, although the moment that sticks out in my mind more was seeing the gray-haired (and bearded) gentleman in the elevator who had a briefcase with a bumper sticker saying "The Opposite of Progress is Congress." Now there's someone who either subscribes to the idea that you catch more flies with vinegar than honey, or actually works in the building. The subtle things struck me the most, however, like other lobbyists showing up at an office, with the staff greeting them very warmly. My guess is that the lobbyist I saw had visited the office several times.

ProSpace has run into problems in the form of rumors that it its platforms support one aerospace company over others. Since then, the participants in March Storm have had to pay a fee to participate in order to cover expenses for training rooms and the like. They continue with their March Storms, and send out occasional updates to members about recent legislative action that may have had help from the previous March Storm.

Nuclear Groups

These groups are typically pretty quiet, unless there's an upcoming launch that involves some form of nuclear material, in which case they are against the activity. The most recent launch that caught the attention of this crowd was the Cassini mission sent on its way to Saturn. This mission had 33 kg of plutonium on board to provide power and heat for the spacecraft to function.

In 1997 when *Cassini* was preparing for launch, The Global Network Against Weapons & Nuclear Power in Space launched a number of legal challenges to the mission. The argument was that NASA was endangering people by launching a nuclear-powered spacecraft, all the while ignoring the possibilities of powering the craft by other means like solar energy. The first part we'll talk about in a bit, while the second is dealt with in Chapter 11, "The Technical Stuff." As happens in all of these cases, the courts heard their arguments but disagreed with them, clearing *Cassini* for launch.

The nuclear activists were not finished. Because *Cassini* was such a huge spacecraft, it couldn't fly directly to Saturn. It required a few flybys of other planets such as Venus (2 times), Earth and Jupiter. More can be found on flybys and the slingshot effect in Chapter 11. The approach to Earth provided a new opportunity to challenge this nuclear-laden spacecraft. Here, their point was that the slightest error in the craft's flightpath would cause it to crash into Earth, spreading its nuclear fuel across the surface. Actually, the case of a wrongly approaching spacecraft is more likely to spread nuclear material than a launch because the craft is moving so much faster as it passes Earth than it does when it's leaving Earth. In this case, though, NASA demonstrated that it had a very good idea of where the craft was going and knew for a fact that it would not impact the Earth. They were allowed to carry out their flyby, and *Cassini* arrived at Saturn in July of 2004.

I find it hard to write about such groups without invoking some sarcasm. While I understand that there are groups of people who will protest anything, in my opinion, this group plays on people's fears by manipulating a few facts. There are many organizations that can be accused of doing the same thing, but this particular fear-mongering rubs me the wrong way, so bear with me.

The two arguments that I've heard the most against nuclear power in space is that it's designed to prepare people for nuclear weapons in space and that any accident would cause a massive release of nuclear material that would poison or radioactively contaminate everyone in the world. Let's look at each argument separately.

The first argument is one that appeals to a number of people with an anti-military mindset, but is not useful at all in a court of law. Consequently, you'll see it more in literature and websites than in any sort of formal documents. The idea goes that world governments want to put nuclear weapons in space, and that they're looking for a way to break it to the public gently. The argument follows that after NASA launches a couple nuclear-powered craft like *Cassini*, then moves on to a nuclear reactor-powered space ships. From there, the Air Force would move in and launch nuclear missiles into orbit. There are a lot of flaws with this theory, but I'll only deal with a couple. First, if the government wanted to have nuclear missiles in space, it would do so. If the mission was deemed too controversial, they could simply classify the payload and keep the whole thing under wraps. A significant number of launches take place today with little or no acknowledgment from the government; adding a few more would not be a big deal. Second, nuclear missiles in space are impractical. The nuclear missiles that the United States has in its arsenal today can be delivered in 30 minutes anywhere around the globe. The additional cost required to base such missiles in outer space is completely outrageous compared to any benefit that would come from the action. Missiles on the ground can be hardened, placed behind concrete doors so thick that anything but a direct nuclear strike won't be able to breach them. Missiles in space cannot have such protection, and the fact that orbits are regular (predictable to a fault) make them an easy target. So the "preparing people for nukes in space" is not a rational argument, but arguments don't have to be rational to gain followers, they just have to be loud.

The second argument is that the materials used within the Cassini mission are both toxic and radioactive enough to kill everyone on the planet, and that an accident would spread the contaminants around the globe. I will concede the first point, but the distortion kicks in on the second. The fact that the first point is true may surprise people, but if the plutonium on board the *Cassini* spacecraft were split into billions of pieces, and one of each of those pieces were placed inside the lungs of everyone on the planet, everyone on the planet would die. How's that for an alarming statistic? The problem with the argument comes in where the plutonium is distributed around the world by a launch accident. When discussing the plutonium on board the Cassini mission, we're talking about material with the consistency of a heavy brick. This material is extremely tough, and even though rocket explosions are spectacular events, they don't transmit much of their energy forward to the spacecraft on top of them. In 1997, a rocket exploded carrying a global positioning satellite, and the satellite stayed in good enough condition to continue transmitting a signal until it hit the water. So it's likely that a rocket explosion would not break the plutonium at all, but even if that happened the bricks would only break into a couple pieces, not into the ultra-fine dust required to "infect" all the people on Earth as described by activists.

As I said, all the controversy over the Cassini mission was for naught, and the anti-nuclear groups have gone into hiding for a while. I'm sure they'll be back, however, especially since the recently announced new direction for NASA will rely on nuclear power for continuous electrical power and heat on board crewed spacecraft.

The International Committee Against Mars Sample Return

In yet another vein of anti-space activities is the ICAMSR. This group formed in answer to early plans for a Mars sample return mission that was originally scheduled to fly in the year 2005. In the days before this project got put on indefinite hold, increasing budget pressure had the team developing the return vehicle coming up with more and more radical methods of cost savings. Towards that end, since in space projects mass equals money (well, time equals money too, in that a project delayed is a project made more expensive), engineers were working to shave every kilogram they could off of the return capsule that would bring back the Martian samples. After making the capsule as light as they could, they decided to try a novel method of landing the samples on Earth…known as a crash. In the plan, the capsule would be a ball holding small portions of the Mars surface, and it would enter Earth's atmosphere at high speed, slow down to about 400 kilometers per hour and smash into the Utah desert without a parachute. Founding members of ICASMR felt that this approach was absolutely irresponsible, seeing that it's possible (the probability is under much debate) for Mars to harbor life forms today. With the ease that the internet provides in creating an organization, they got a website and created one.

Of course, most of this is less urgent right now, since the sample return mission that they were against has been delayed beyond the foreseeable future. It's likely, however, that if Mars takes on its expected new priority today in the post-*Columbia* era, we'll be hearing from ICAMSR again. Maybe they'll have a more user-friendly acronym.

UNITY: THE ANSWER?

Having volunteered in the space advocacy arena for a few years now, and after witnessing one of the splits that took place within it (Mars Society and Mars Institute), it bothers me that people of similar mindset will split their efforts over minor differences instead of unifying their efforts for what they have in common. Anyone with a specific agenda in mind for space travel's mind is racing right now, thinking such things as "Mars is the only destination that makes sense" or "we have to turn near space over to private industry and have government explore the far frontier" or other themes appropriate for their organization. So, when I ask whether those differences matter, the answer from most of the space advocacy community will be an emphatic "YES," but to the public at large, I believe the answer may be "no."

The problem comes in that combining messages waters all of them down. By the time you mix the messages of all the societies mentioned above, things boil down to "go into space," which, even though it's short enough to make a useful slogan, lacks a certain oomph, shall we say. Immediately after the *Columbia* tragedy, I contacted the leader of one of these organizations, and asked if there was any interest in putting together some sort of combined message from all the advocacy groups. The response was lukewarm, and there was no combined message.

What if an organization formed whose only requirement for membership was a membership in any of the above organizations? That new organization, claiming more members than any other space advocacy organization, and might be able to pull together a coherent, useful message.

In Senate hearings late in October 2003, high-ranking people from three space organizations were asked to help form a coherent vision for NASA's future. The Space Frontier Foundation, The Mars Society, and The Planetary Society speakers all agreed that Mars was the place to go for the NASA's next goal. So, in some ways, progress is being made.

Today, it's unclear what the overall space activism response will be to the new space policy, though some have already come out to say that The President should have spelled out a more active role for commercial interests.

Chapter Six

A Robust, Expanding Space Presence

OK, we've discussed space's past. We've talked about some of the myths attributed to spaceflight, and we've talked about many of the current space efforts that are underway. While I've tried to maintain a level tone, I don't believe that the current level of space activity is neither high enough to maintain a significant inspiration for children, nor is it moving forward at any pace that will be looked upon as good in the future. I'm the kind of person who doesn't like to describe something as a problem unless I also have a solution to offer, but what is the solution here? In order to get where we need to be (on the track to being a solar-system wide civilization), people must form a robust, expanding space presence. Here are some of what's required to do that.

Multiple Methods to Orbit

If you need to travel from the East Coast to the West Coast of the United States, you have several options available to you. Most likely, you will purchase a plane ticket. Hopefully, you'll be able to take advantage of some sort of sale and get a good price, but you'll be able to purchase the travel for a relatively small part of what you earn for a year. This simple action pulls many factors into action that you may not even think of. There are several airline carriers, and you can choose between them based on cost, convenience, in-flight service, or even based on some previous event where you were treated badly. You have a choice. If, for some reason, the airlines aren't flying or you choose not to fly, you have other choices. Trains provide a slower pace, but will more than likely get you to where you want to go, with a chance to see everything along the way (at least what's near the tracks).

Depending on the train service available, it may actually be faster to drive yourself across the nation instead of taking a train. Sleeping in your car cuts costs even more. Of course, last, and likely least in overall popularity, is bus service.

The point is that the number of choices in methods to cross the country is mind-blowing. This choice is the power of a consumer-based economy, and it is completely absent in space travel. Some will argue that there is a choice in methods to get to space. Two men named Dennis Tito and Mark Shuttleworth could not pay for a ride into Earth orbit through the United States space program, so instead they paid Russia for a ride. I think that the only choice they had was to fly with Russia or not fly at all. Even though some notable people have proposed flying paying civilians on board a space shuttle, (Buzz Aldrin, a crew member of the first moon landing mission, is one such notable) the prospect hasn't been taken seriously.

So, if you are a "bona-fide" space traveler, not one with the means to pay their way, and you want to get into space today, you have two options: The Space Shuttle (when it starts flying again) and the Russian spacecraft *Soyuz*. If you're going to fly on *Soyuz*, you'd better pack a small suitcase, because the carry-on bag requirement for a *Soyuz* capsule makes carry-on limits for airliners look pretty large! If you have any checked baggage, it will probably have to travel on another spacecraft. Oh, and by the way, you'll have to give up 6 months of your life to travel to Russia for training in speaking and reading Russian, as well as learning emergency procedures so that you can help out in case of a problem. For those of you who've flown in an exit row on board an airliner, in *Soyuz* you don't have the option of swapping with someone else if you feel you can't perform the duties required.

Those two options are it, and that's only if you've passed the rigorous application process, medical screening, and physical training as we've described before. If you're just someone off the street, your odds are pretty slim.

When the *Columbia* broke apart over the Texas skies in February of 2003, a critical portion of the plans for building the International Space Station went with it. With the shuttle out of action, the only way to reach the ISS is through a Russian flight on a *Soyuz* vehicle. Unfortunately, the *Soyuz* can't take enough supplies to the ISS to keep it functioning at its full crew of three. Until the shuttle is flying again, the station will have a maximum of two people, and may have to be abandoned for a time.

This situation is completely amazing. Imagine a building being built in a city, when suddenly construction had to stop because a crane like the one building that building broke. Space travel is the only effort run in this manner.

When an airplane crashes, are there immediate calls to ground the entire fleet? No. There was one exception, in the supersonic Concorde aircraft that crashed in 2000. The powers-that-be controlling that plane made a supreme effort to get the remaining planes flying again, only to stop flying the aircraft in 2003. Even in the extreme case of multiple crashes of one type of aircraft in a short period of time, where a grounding of its siblings is warranted, the overall impact to the airline industry is manageable. This is again because of multiple options available.

Imagine something like the *Columbia* tragedy taking place in a different world. Imagine if an assortment of spacecraft of similar capabilities to *Columbia* flew every day, taking people or materials into orbit. Perhaps the different spacecraft that fly are not the same. Some launch from sea platforms while others are dropped from airplanes. Some take off the old-fashioned way with a liftoff, but they'd all have something in common—a lack of focus on the takeoff EVENT. People don't cheer every time an aircraft rolls down the runway and takes flight, so they're free to focus on the results of the flight. This is the type of world where space travel can play a major role in everyday life.

In this alternate world, the loss of the craft (and maybe a crew—one of the advantages of having multiple methods to go to and return from orbit is that not all the spacecraft would need to be crewed) would be an inconvenience. If a grounding of that type of ship was necessary, payloads would have to switch to another spacecraft for their ride into orbit. It's possible that some craft of the same design would still be in orbit (remember, we're in an alternate world where there's more than one spacecraft flying at a time, and no one notices) and would need to be checked out based on the accident, but people and goods could continue to move between the Earth and space.

Instead, we're on our world. The disintegration of the *Columbia* has grounded the entire fleet of orbiters while national-scale investigations work to determine what happened. Most portions of government have gotten involved in the inquisition, and the political posturing started before the report on the accident was even released. Somewhat lost in all this is that there is no way to lift heavy payloads to the space station during this search for answers and solutions. There are craft that can lift enough mass to the station that they could do some heavy lifting in place of the shuttle, but the pieces of the station were designed to fly on the shuttle alone. There's either no way to connect them to the other rockets, or the other rockets shake in a way that's a little different than the shuttle, and would cause the station pieces to shake apart.

So, we wait.

SPACE: WHAT NOW?

TWO (AND A HALF?) REUSABLE STAGES TO LOW EARTH ORBIT

Most rockets today use two stages to get to LEO. In these cases, the first stage is dropped off fairly early in flight, and the second enters orbit with the payload, usually falling out of orbit in a few weeks, as the payload uses its own thrusters to raise its altitude and stay for a while. Others use three stages, where the first gives an initial push, then drops back to Earth, the second accelerates the payload closer to orbital velocity, but not quite there, then it also drops back to Earth. The third stage gives a substantial push to keep the craft in orbit. In most of these cases, the third stage stays connected to the satellite.

Here, the proposal is to use three total stages as just described, but have the first two stages be reusable. For a flight example, let's assume that the origination point is Florida. The first stage, dropped fairly early in the flight profile, but at a pretty significant altitude, glides back to the launch point. This design is currently favored by NASA for its flyback shuttle boosters to replace the current solid rocket boosters, although there is question as to whether or not this system will be built given the re-examination of the shuttle program after *Columbia*. The second stage, after doing its job, drops away and continues ballistically to the top of its arc, then falls through the atmosphere and parachutes to a landing on the other side of the Atlantic. This stage could then be loaded into a standard cargo container and shipped back to the US. The third stage, pushing our payload into orbit, is the most critical as far as mass goes, because all of its dry mass goes into orbit. With the idea of eventually using the architecture for human-crewed flight, this third stage has two firing modes. The first one, used in emergencies to get the crew away from a failing booster, fires with maximum power for a relatively short period of time. This will accelerate the crew away from disaster, to a safe landing somewhere ahead. In most cases, this mode won't be necessary, and the stage will fire in a lower thrust mode over a longer period of time to put the payload into its final orbit.

The big mindset change for this system is the use of commercial shipping to return the second stage. Other proposals for the first and third stage have been brought up before in one way or another, but the idea of a downrange recovery of a second stage, it being placed into a cargo container and shipped, is new. The idea came in an attempt to decrease infrastructure, while at the same time trying to minimize complexity for the second stage. Other proposals bring back their second stage, but Kistler Aerospace, for example, plans to return its second stage after taking it the whole way to orbit. This is certainly feasible, but it also requires much more in the way of heat shielding in order to bring the stage back intact. A second stage that falls back to Earth at something less than orbital velocity will not have to be as well protected.

The somewhat long turnaround process for the second stage (on the order of weeks, though I'm sure containerized shipping could make it much faster if the landing zone were located near a port) would require a rather large number of these second stages in order to maintain a high launch rate. This requirement for many copies of the stage will result in an initially large production run, lowering per-unit cost.

The first stage will be able to turn around and be ready for flight much faster than the second, since it returns to the launch site on the order of an hour after launch. Therefore, there won't be a need for as many first stages as second stages, which could cause higher production costs. The best balance to strike between production levels and cost will rely on someone else's knowledge, but also on market forces setting the number needed for requirements at the time.

Re-use of rocket stages this often will bring space operations into the realm of commercial aviation. With that transition comes some unwanted effects, including a new concern for metal fatigue. Metal fatigue made headlines early in the days of passenger jet flight, as the British Comet suffered a number of failures and ended up falling out of favor because of them. Boeing, a new competitor on the world stage at the time, took advantage of the Comet's misfortune and sold a huge number of its 707 aircraft. Metal fatigue comes about when a component of an aircraft or rocket is exposed to many cycles, or times of having stress put on it, and then released. The fact that spacecraft are designed and built with very little additional metal on them will not work to help in this situation. Another problem that rocket builders will come across is their work with extremely cold (cryogenic) liquids. Repeatedly exposing a metal part to these kind of temperatures while stressing it will only make the condition worse.

STAND-BY SPACE RESCUE/ORBITAL SAFE HAVENS

Rand Simberg, a space entrepreneur and web-based pundit for space issues, compares our sending space shuttles into orbits not occupied by the ISS today with sending people into the wilderness with no hope of rescue[60]. To use a modern-day, ground based comparison, imagine driving off the in the desert in your sport utility vehicle, not having told anyone where you were going, nor carrying any phone or navigation device with you. If your vehicle breaks down, you are out of luck. This example is a little extreme, but it's necessary to illustrate how alone the crew of *Columbia* (or any shuttle mission that doesn't rendezvous with the space station) really was during their mission. Yes they had communications all day and all night with the ground. Yes they took amazing images of the Earth and the moon and beamed them back to our planet for people to gawk at, but when it came time to button up and come home, they had no way of knowing that their spacecraft was hopelessly damaged. Even if the crew did know, I still believe the risks involved in trying to save the *Columbia* crew have not been completely understood. This condition doesn't need to remain, however.

Stand-by space rescue is not entirely new to the United States' space agency. During the Skylab missions of 1973-74, when one crew experienced problems with their return craft, efforts went into preparing the next scheduled mission as a rescue mission. The problem with the return craft turned out to be isolatable, so the rescue mission was not required, but the rescue plan would have likely worked. Skylab was unique in US space history, however, in that the crew in question was scheduled to stay in orbit for 56 days, giving teams on the ground plenty of time to get the rescue mission together. Supplies existed on board the station to allow an even longer wait if necessary.

The space shuttle is limited in its time it can spend on orbit, though. The primary limiting factor is the hydrogen and oxygen it uses to make electricity and drinking water for the crew. While it's possible for a shuttle to stay on orbit for up to a month, that sort of thing needs to be decided before launch, so that additional supplies can be placed on board to allow the mission. If a shuttle travels to the ISS, the crew's ability to stay on orbit is increased because they can relocate to the station. Waiting beyond a certain timeframe, however, would likely doom the shuttle, because once its supplies of power were used up, there are no procedures for refueling the craft in space.

One answer to this problem is to have a standby space rescue service ready to launch on a couple days' notice to rescue a crew in orbit. Please note that when we achieve the goal described in the first section of this chapter, multiple methods to orbit, the requirement of standby rescue should be met. (This assumes that all the available spacecraft are designed with some compatibility in mind, and setting those standards is an excellent role for a government agency, by the way.) In modern-day terms, this translates to having a space shuttle sitting on the ground that can be made ready to launch in a fewer number of days than the scheduled mission duration. Granted, if the scheduled mission were headed to the International Space Station, ground crews would have a little more time to prepare. The question is: is this feasible?

It is possible for the space shuttle to fly with only two people. The first four shuttle missions did just that. Since most shuttle missions today carry seven crewmembers (note the ease of using the terms astronauts or cosmonauts is gone, because many crews today are made up of a mix of nationalities), the inside of the shuttle will have to be changed to handle the crew total of nine returning. Once in orbit, this sparsely crewed spacecraft would rendezvous with (get near) the stricken shuttle (or station) and commence rescue operations.

Here, things get a little sticky. If the broken shuttle is docked with the space station, the broken shuttle will have to move away from its berth to allow the rescue shuttle to dock. Before doing this, the rescue shuttle and mission control will have to decide whether or not the broken shuttle is fixable. If it is, the decision will have to be made as to whether or not to fix it while attached to the station or to separate the craft first. In order to conduct repairs, its likely that the "good" shuttle will be grappled with the space station's arm and held in place, and then be used to repair the stricken shuttle. If and when repairs are complete, the crew can split up however they like to ride home in the separate shuttles. Personally, I'd ride in the one that wasn't repaired on orbit.

If the shuttle in question is not fixable, things are a little easier physically. In this case, the crippled space shuttle is simply separated from the station and allowed to drift away. If it still has power, ground controls can fire the rockets to return it to the atmosphere, where it will burn up with its pieces falling into the ocean. The rescue shuttle then docks with the station and brings the crew of the first shuttle home. Of course, then the world has to face continuing the space station program with only two shuttles instead of three.

I spelled out the previous scenarios to show that they were possible. In reality, I don't really think that they are practical. Even if a stand-by shuttle becomes a requirement for future shuttle missions, I think that as operations resume the definition of "stand by" will change. For example, if the requirement reads that a backup shuttle must be on the pad before another shuttle flies, reality will make NASA slowly push more and more processing to the pad, so that the craft's being there means less and less as far as a stand-by goes. It's likely that a shuttle that's on the pad for a station assembly mission will require some pretty serious payload changeout before it would be ready for a rescue mission, and that would delay the rescue launch as well. NASA recently stated that a stand-by shuttle will be ready for launch whenever another is flying. There have been some half-hearted attempts to try and rectify this situation, however.

NASA considered the possibility of creating another system that could deliver supplies to the ISS. The Alternate Access to Space (AAS) program was designed to allow low-cost delivery service to the ISS. Essentially, NASA would release a contract saying "We need 2000kg of supplies delivered to the space station." By having this cargo be low-cost items such as food, water, or fuel, the theory was that a new rocket builder could cut its teeth on the new payload requirement, and that the loss of one or two missions occasionally wouldn't be catastrophic. After all, if 2000kg of water drops into the Atlantic Ocean, it's pretty easy to come up with 2000kg more. This approach is quite a bit different than normal government contracts, where pages (in many cases, hundreds or thousands of pages) are devoted to how a company that wanted to win the competition should carry out the activity that NASA had in mind. The simpler contract was designed to appeal to entrepreneurs, people who had a new idea of how to get into space but couldn't find investors in their project. With this guaranteed contract for delivery-on-orbit, these start-up companies could make a better case to investors, and get the funding they need. All this would take place at a much lower than usual cost to the government.

Unfortunately, this program ran into trouble, and part of it came from a bizarre concern in my opinion. First off, the more legitimate concern was that NASA didn't want someone's home-built spacecraft approaching the space station. Here, the worry was that the spacecraft was not completely thought through, built by monkeys, or something along those lines. This being the case, the craft would fly into orbit with no trouble, maneuver to the space station's orbit, and then break down on final approach. Unguided, it would crash into the station and damage it. Something similar happened to the Russian space station Mir. If my tone makes it sound like I don't think this is a very good reason, my efforts at conveying my opinion were successful. Yes, orbital collisions are bad, but if there's a concern about approach, then NASA should build the cargo craft and let a contract to deliver it to orbit. Ironically, that's the kind of solution that was proposed to the other problem, which was even stranger.

Somewhere within NASA, someone decided that any spacecraft docking with the ISS could not have any advertising logos on it. The reason why this was important was never released, but it put the Alternate Access to Space program into serious trouble. How many delivery trucks, trains, or aircraft can you think of that don't have advertising logos on the side? In an attempt to save the AAS, some people proposed a small station (presumably built by

NASA, at an undisclosed cost) that traveled with the ISS in orbit. Delivery vehicles would travel to the small station and unload their cargo. Then, another spacecraft (again, presumably built by NASA, at an undisclosed cost, and with no advertising logos on the side) would take the cargo to the ISS. So, because of a fear of advertising, a potentially cheap method of delivering non-critical cargo directly to the space station became an expensive, multi-step process that likely won't ever happen. There has been some recent interest in bringing this type of program back again, and personally, I have to give Alternate Access credit for one thing—it served as partial inspiration for my version of the Orbital Supply Depots program discussed next.

On-Orbit Supplies

When a spacecraft launches from Earth to travel beyond low Earth orbit, most people would be amazed to learn the small percentage of mass that actually flies beyond LEO. Of course, most people would also be surprised to find out how little mass actually makes it into LEO. For example, the Mars Exploration Rovers launched in June and July of 2003 weighed 185kg each. That mass is just the rover itself, but there are plenty of other parts that need to leave LEO in order to get those landers on the surface. There's the pyramid-shaped cover surrounding each rover, airbags covering the pyramid, and a heatshield/backshell combination to protect them from the high-speed trip through Mars' atmosphere. A cruise stage is attached to the backshell for most of the ride to Mars, which provides the entire craft solar power, and allows it to change directions on the trip if necessary. In total, 1063kg had to leave Earth orbit to get 185kg of rover on the surface of Mars. Yet another multiple comes in when considering how much the rocket weighed that lifted the Mars craft off the Earth.

Still, the physics of getting mass into orbit are understood pretty well. Where we fail in cost is our execution of the physics. Several years ago, Arthur C. Clarke pointed out that the pure energy cost of putting a human into space is $100, and that any cost above that is our own incompetence. But how do we increase our competence?

The Plan

Here, the goal is to provide a reason for frequent flights into Low Earth Orbit (LEO). To make sure that the flights are not frivolous, each is tasked with taking a significant amount of raw material (water, or its components, in this case) to a destination in orbit. By allowing an arbitrarily high flight rate, with relatively low costs for the loss of a payload, a startup company can build its expertise in rapid-launch rocketry. After flying a number of missions (let's say one a week for a year) they have a sound statistical case for claiming that their vehicle is safe to fly. Then, that vehicle could be used for other applications in LEO such as space tourism. But first, we need a destination.

An Orbiting Supply Depot

Imagine a place in Earth orbit where any craft in that orbit could stop and pick up hydrogen for use in rockets or fuel cells, water for its crew, and oxygen for multiple uses. For missions to LEO, these supplies could cut the need for frequent logistics flights from Earth's surface. For missions beyond LEO, these supplies could make such journeys feasible using the relatively modest-sized launchers that we have today.

The entire system can be described as an orbiting electrolysis station, where moderate (10,000 kg class) payloads of water are delivered and converted into liquid hydrogen and oxygen. Since water can be stored on orbit with relatively little maintenance, there is no penalty for a high flight rate, as the water can simply be held in place until the electrolysis process catches up. Visitors can then take on the supplies, as required. This type of station has been proposed before[61], but its implementation was seen as being feasible after some sort of mass-driver exists. I assert that waiting for a mass-driver moves the propellant depot into a realm of super-science, which is unnecessary to make a supply depot useful.

The Depot

The depot is designed to fly in a single flight of a heavy-class payload by today's standards, which translates to Evolved Expendable Launch Vehicle, Shuttle, or Ariane V. If constrained to a shuttle launch by either politics or deployment complexity, the cost numbers discussed later will have to be increased. At one end of the station reside the docking units for water deliveries. The system is designed to allow multiple dockings at once. From these docking locations, the water flows into an electrolysis unit, which splits the water into its components: hydrogen and oxygen. From there, the component gases move to a refrigeration system that cools and pressurizes them into their liquid form. The cryogens are then forced into tanks for storage by the gasses produced behind them.

At the opposite end of the station rests the docking unit for customers, as well as a set of thrusters to provide acceleration as needed to make all the fluids flow in the proper direction. Acceleration along a primary axis allows a simple flow method that doesn't require pumps, bladders, or any of the other contraptions that come with moving liquids in zero G, providing for a much simpler system. The docking unit allows access to the liquid hydrogen and liquid oxygen from the cryogenic tanks, as well as to the water from the original deliveries.

My original goal in this design was to stabilize the station passively, though as the idea matures, that becomes less likely. Because of the huge power requirements for both the electrolysis and the cryogenation process, any solar arrays supporting the mission would have to be larger than that of the ISS. By my estimation, it would be best to hold them steady and pointed at the sun by turning the entire station. This has the side benefit of keeping the cryogenic tanks in the solar arrays' shade.

If the system catches on, and customers need supplies at a greater rate than one depot can produce or store, expansion is a possibility. By docking an additional station at the customer end, it's possible for the first station to double its storage capacity. Increasing production rate is another matter, because more power will be required, and that would be difficult to accomplish with solar power, considering that the entire second depot would be shaded by the first station's arrays.

I foresee the depot being built as a typical government procurement, with all the trappings of budget overruns, congressional hearings and news-show installments. Once completed and on orbit, it will be time to pay someone to deliver water to the station using capsules built largely the same way.

The Delivery Capsule

The capsule that carries water to the depot will be designed around a tank with docking ports at each end. Being dual-ended allows capsules to dock either to the depot, if they're the first to arrive, or another capsule if it arrives later. The tank will be insulated and heated as necessary to keep the water onboard from freezing. Power for the heaters will come from the depot when the tank is docked. The tanks are designed to handle pressure, as waste gas from the fuel depot will be used to pressurize them, and allow a rapid flow of water into the system. As tanks empty, the same pressure will force them to disconnect from the depot when the tank is unlocked. These tanks may become useful in future space projects, serving as storage areas or perhaps even living quarters, assuming that they are properly proportioned.

Delivery to the depot from the injection orbit, as well as proximity and docking maneuvers, are handled by the orbit assist ring or OAR. The OAR provides enough propulsive power to move the capsule to a docking with the depot, keeping the water warm as necessary. Once the capsule is in position, the OAR may separate to de-orbit and burn up in the atmosphere, or it could be used to move an empty tank to a new site for use.

Depending on the method chosen to liquefy the hydrogen and oxygen, it may be necessary to have additional cargo vehicles bring liquid helium to the depot. This complicates the system somewhat, but the trade may pay off in a simpler, longer-lasting supply depot. It would also increase the flight rate necessary, if helium were expended in the process.

In this discussion, the lift mass of the water is more important than the lift mass of the delivery capsule, but the capsule's mass is non-zero, so it needs to be discussed. Estimates on the mass of the capsule and the OAR range from 15-20% of the cargo carried, so any discussion of launching 10,000 kg of water into orbit will need to include 2000 kg for the tank and OAR.

The Launch Vehicle

In this concept, the design of the launch vehicle is of no consequence to the outcome. Because of the low payoff per kilogram offered and the high flight rate, it's likely that the vehicle and its support operations will have to be radically different than any method so far devised to fly into LEO. I believe that a significant portion of the vehicle will have to be reusable to prevent the complete manufacturing of each one for each flight. Because of the high cost of labor in the aerospace industry, it's likely that the production and launch teams for this vehicle will be small. By taking both actions, the primary flight cost will be propellants, not wages. What makes this transition possible is payment upon delivery of the water, not payment for booster development, as is the norm today.

The delivery-on-orbit contract is becoming more common in government space activities, with examples including the Navy Ultra-high Frequency Follow-On (UFO) communications satellite and the Geostationary Operational Environmental Satellite (GOES) series due to fly in the next years. The version of this method envisioned here would set a new precedent. Previous delivery-on-orbit missions were small scale, on the order of 5 total launches. These kinds of numbers never allowed a real cost savings because by the time a team got used to the delivery-on-orbit approach, the mission was over.

For this method of delivery-on-orbit, the numbers are much larger. After choosing a payload size (10,000 kg used here), a dollar amount for that launch is chosen. For this example, we'll use $10M per launch, yielding an unheard-of per-kilogram rate of $1000. A flight rate must also be chosen. Here, I'm partial to one a week for a year, because it's an unprecedented flight rate in modern times, and any vehicle that can fly once a week should be able to fly more often, as well. If this rate is chosen and guaranteed for one year of flights, it would provide an incentive of $520M to a company who has a 100% success rate on their vehicles.

The Payoff

At the end of the initial round of 52 flights, figuring a 90% overall success rate, there would be 468,000 kg of supplies in LEO, stored as either water, hydrogen or oxygen. This is an amazing amount of mass, and it would allow a lot of actions that were unthinkable before. Assuming a total cost of $1B (see cost estimates in the next section), the dollars spent to get a kilogram of this useful material into orbit is around $2150. If no one else has adopted the cargo launcher as theirs, and flights are still taking place using current expendable rockets and/or expensive shuttles/replacements, these supplies can be sold at a profit in LEO.

If, on the other hand, the booster used to loft water to the depot becomes the booster of choice (a pretty big assumption, involving true market forces coming into play in the space industry, but work with it), then anyone using the low-cost booster will not need to purchase supplies on orbit. Here, the goal of low-cost flight to LEO has been achieved, and the supplies at our depot can be used for deep space exploration. For example, the Atlas III booster can launch 8,610 kg into LEO using its Centaur upper stage. Assuming that the Centaur were upgraded to allow precision maneuvers and if that stage were refueled in orbit then allowed to fire again, it could send its payload on the same trajectory that the Cassini mission traveled in its journey to Saturn. The costs of flying on a Titan IV rocket (what *Cassini* actually flew on) vs an Atlas would be the maximum savings recorded using the depot. In another example that I find more exciting, the Mars Direct mission architecture for sending people to Mars requires two launches of 140 tonnes into LEO. About 100 tonnes of that LEO mass is propellant. So, with a propellant depot on orbit, only 40 tonnes of crew habitat and empty tankage would have to be launched from Earth, which is feasible with the boosters of today.

The Cost

I believe that this plan can be implemented for a cost on the order of $1B. This includes 52 delivery missions to orbit at $10M each, building the supply depot for $250M, launching the depot on a heavy EELV for $150M (a planning number—individual flights are negotiated), and building 52 delivery tanks for $100M. This is not a small amount of money, but given the payoffs already discussed, it is money much better spent than that thrown down a dead-end research project.

The costs mentioned above are only for a successful project. If, for some reason, no carriers rise to the challenge of taking deliveries of water to an orbiting depot, the project can be abandoned at a significantly lower cost ($350M assuming that the depot and all of the delivery tanks are built) without launching anything. In the case of a partially successful project, only deliveries are paid for, so there are fewer supplies on orbit but at a significantly reduced cost.

It's possible that $10M for 10,000 kg is a completely unrealistic cost for launch to orbit. If this turns out to be the case, and the goal is still seen as valid, the price can be raised to provide further enticement.

On the other hand, there may be more interest in meeting the challenge. If more than one company builds a viable vehicle (demonstrated through flights, of course, not just presentations) additional carriers and depots may be built, all the while increasing our resources in orbit.

A Demonstrator Mission

A lot of things required for this mission are relatively common on Earth, but haven't been done in space. A demonstrator mission that flies an electrolyzer and cryogenic cooler would go a long way to answering some of these questions. A rather simple demonstrator payload could fly on board the Shuttle or Station if it meets safety requirements, but this demonstrator would have to rely on some sort of microgravity system pushing fluids through. If the mission flies on its own, it would provide a more rigorous test of the system, including the simplifying acceleration/pressure feed described earlier. The freeflyer demonstrator mission would be much more complex and expensive, but would model the idea with much more fidelity.

The Issues

Unfortunately, this system is not ready to go, out-of-the-box. There are issues—from the technical to those dealing with established attitudes—that must be overcome before this concept can become reality.

The first is power. This depot would require a huge amount of power, first to electrolyze the water, and second to cool and compress the products. Even with unprecedented power supplies (100 kW) devoted solely to electrolysis, the water processing rate would be only 8.18 kg per hour using an existing ground-based system. At this rate, it would take 51 days to process a 10,000 kg load of water. Of course, the electrolysis process is very scalable, so more power would allow a greater flow rate. Power questions may turn the system on its head, requiring delivery of liquid hydrogen and oxygen to the station instead of water. The hydrogen could be used to cool both liquids, then to provide power, water, and fuel for the depot's engines. This design would require much less power. The type of power to use (solar vs. nuclear) is a favorite topic for some, and if nuclear power becomes more acceptable through NASA's Nuclear Initiative, it may be an option, but that's a discussion for another day.

Once the products are generated, storing them is not easy. Hydrogen is especially difficult to keep, due to its low density and notorious tendency to leak. Adding additional depots below the first can solve the volume problem, but this won't be practical until the concept has proven itself useful. Leaks in valves and pipe joints will have to be minimized through meticulous construction.

The supply depot will be one of the oddest spacecraft ever built, as far as attitude control. It will start out relatively light, with large solar arrays and the majority of its weight on one side of them. As water deliveries accumulate, the center of mass on board will shift radically, only to be offset slightly by processing through the electrolyzer.

In real estate, the three most important words are location, location, location. In orbit, this translates to inclination, inclination, inclination. Where should an orbiting supply depot go in orbit? The answer is not intuitively obvious. The space station is inclined at 52 degrees to the Earth's equator, but the most efficient launches from the United States take place at around 28 degrees. Any corporate interest worth its salt would want the largest advantage possible in its launch, and base operations near the equator. This will require some serious marketing analysis, the likes of which space activities have never seen before.

The fact that nothing like this has ever been done is always a momentum barrier. A good answer is, "That's OK, low-cost launch to orbit hasn't been done either; maybe they match." Another philosophical point that needs to be addressed is that this system is biased towards hydrogen-oxygen systems. It's true that the electrolysis process produces those propellants in excellent proportion for a H2/O2 engine, with a convenient remainder of oxygen for use as breathing air, but a nuclear-thermal rocket could easily stop by for a fill-up, taking only hydrogen for its engine. If the payload for this rocket were a crewed capsule, they'd still be able to use the oxygen and water that the system supplies. Left-over oxygen could be sold to orbital stations with other crews on board. Let's face it; humans are addicted to oxygen, and without some major genetic alteration, that isn't going to change just because we have stations in orbit.

A Utility Spacecraft

What comes to mind when you think of the word utility? It's likely that city dwellers will think of water, gas, electric, and other things that, for the most part, you don't have to think about in life—they're just there when you need them. Others will think of the usability of an item. How many jobs can it do? How well? We need to start thinking the same way in space.

Since the beginning of the space age, at least in the United States, when a new mission was envisioned, a new vehicle was built. Granted, the sample set here is extremely small, and things did get a little better as time went on as far as utility (the space shuttle, for example, can launch satellites or become a small space research laboratory, or deliver pieces of the space station along with a construction crew). But when the US space program was charged with landing someone on the moon, they built two new spacecraft (*Apollo* and its Lunar Module). Along the way, they decided to build another (*Gemini*) to practice some of their plans while the real system was being built. Once the Apollo missions had run their course, when new horizons beckoned, NASA embarked on building another spacecraft (the shuttle).

Now, as the US space agency does some soul searching, it's looking for another space vehicle to take crew to and from the space station. The current term for this vehicle is the Orbital Space Plane or OSP. OSP started as part of the Space Launch Initiative (SLI), and will likely fly on board an existing booster, the Evolved Expendable Launch Vehicle (EELV) run by the Air Force. While the "look" of the OSP is still undecided, initial concepts show it as a small craft described better either as an aircraft (a more correct term is a lifting body, where short wings provide a small part of the lift—the space shuttle is kind of a mix of the two) or a capsule (like the early spacecraft in the US space program). Here, there's likely to be a lot of pressure to avoid "backwards" motion. I believe that any move to build a "capsule" will be viewed as a step back, while a shuttle-type lifting body will be viewed as a step forward. A problem comes in that a capsule-shaped spacecraft will be able to adapt to more missions than a winged vehicle, and long-term usability should be kept in mind for this system, but I digress.

Under the new space initiative proposed in January 2004, the OSP has been cancelled, with a new system named the Crew Exploration Vehicle (CEV) replacing it. I believe that this action was taken to delete the word "Plane" from the concept for our next space vehicle. This frees the system from preconceived notions (like wings) and should lead to a better craft overall.

The bottom line is that, once again, we have a new mission and a whole new design effort.

The time has come to change this approach. Instead of building a new spacecraft for each mission as that mission is assigned, the next series of missions must be determined, and a spacecraft that will meet all of them (with some modifications, to be sure) should be built based on those requirements.

So what are the requirements for such a vehicle?

1. Reconfigurable interior—a utility spacecraft should come off the production line with a minimal amount of mission-unique hardware attached. An unstated part of this requirement that's more of a good idea than a requirement is that the interior be simple. By simple, I mean cylindrical or conical, with mechanical connection points on the interior surfaces that let the inside be changed from a crew carrier (for missions 2, 3 and 4 below) to a long-term living space for a smaller crew (for mission 5 listed below, and any others that come up). Of course, there are plenty of options in between.
2. Reusability is a plus, but not a requirement—because this spacecraft is at the top of a rocket stack, additional mass added to make it reusable cuts directly into the rest of its capabilities. Single-use spacecraft have liabilities, to be sure, but there are other concerns as well. If you're going to build a few reusable craft, how many should you build? If these vehicles were to truly take on their utilitarian role it's possible that you'd have one or more docked at a station at any one time, one on standby for rescue of another craft, one or more assigned to lunar duties, and more assigned to advanced missions. For more information, see Chapter 11, "The Technical Stuff."
3. Provide rescue capability for the shuttle—I listed this requirement next because it's the most politically sensitive right now, and it also has the most muscle behind it as the shuttle returns to flight. A spacecraft configured for shuttle rescue could be maintained in a ready-to-fly condition at all times, stored close to where EELV missions fly from. In case of a problem with an on-orbit shuttle, the payload being readied for flight on the EELV will be delayed, and the crew rescue vehicle will be placed atop the booster. This mission has become less important with current plans to decommission the space shuttle in 2010, although congressional action on the 2005 budget may make the shuttle fly until its replacement is ready.
4. Provide rescue capability for the space station—A spacecraft configured similarly to the one for mission 3 could be launched and stored at the station for a long period of time. It could also serve as a ferry taking crew members to and from the station routinely. In the case of some sort of evacuation requirement, the crew will get on board the craft and return to Earth using it. Because the craft is sized to rescue a shuttle crew of 7, it has the capability to support space station crews of the same size. Evacuation

of the station while a damaged shuttle is docked will increase the passenger load (many times, there are ten people at the space station when the shuttle is docked) and create other problems—if the crew of a damaged shuttle takes the 1 station lifeboat away, how will the space station crew get home if they need to?
5. Provide an alternate route for a crew to get to the station—The wording of this requirement was deliberate to use the word "route" instead of "access" as discussed earlier. If the space shuttle were to be grounded for maintenance problems (as happened in summer of 2002 with cracks in propellant lines) or for another tragedy (*Challenger* or *Columbia* class), this utility vehicle would be available to take crews up to the station and bring them back. It would also have a capability to carry much more cargo than a Soviet *Progress* craft currently used. Once again, conditions have changed since I originally wrote this section. Given the "expiration" date for the shuttle right now, once station assembly is complete, there may be only one route back and forth from the station (the Russian *Soyuz* vehicle) until this utility vehicle is constructed.
6. Provide for long-duration crewed space flight beyond low Earth orbit—This spacecraft, properly designed and built, holds tremendous potential as the basis for our next steps beyond low Earth orbit. If tanks that hold on-board supplies (air, water, fuel, etc) are close to the skin of the spacecraft and can be supplemented by tanks mounted on the outside, the possible mission duration for such a vehicle increases dramatically. This exterior reconfigurability, combined with the interior reconfigurability, makes this vehicle live up to its title as a utility vehicle. Care must be taken here, because trying to do too many things has been a problem in space programs of the past, and this utility vehicle may serve best as a "bailout" capsule. Such a capsule would bring a crew back to Earth from a deep space mission, their having spent most of the time on the trip in a dedicated habitat that housed them comfortably but couldn't bring them back to Earth's surface.

Granted, this approach of trying to do everything can be taken too far, as many say the space shuttle was. The result in that case was a vehicle that did many jobs, but did not particularly excel in any of them. The shuttle can carry cargo and people, but a larger cargo capability would be helpful for building large space structures. The shuttle can't stay up as long as many people would like, and in order to provide a longer-duration space platform, the space station is being built. As for safely carrying people, the shuttle's proven to have some difficulties as well.

Separate Cargo (2 types) and Personnel

Earlier, we talked about creating a cache of supplies on orbit. Here, the discussion turns to the idea that people should fly separately from huge pieces of equipment, and that huge pieces of equipment should fly separately from large shipments of bulk items such as water and food. This idea echoes a recommendation from the Columbia Accident Investigation Board, with the addition of splitting cargo into two types.

The current Space Transportation System, known simply as The Space Shuttle, was designed in an effort to lower the cost of flying people and material into space. In order to increase the number of the vehicle's flights, the shuttle was designed to carry everything that was scheduled to fly into space at that time. As an added insurance policy, Congress made it a requirement that everything on the drawing boards for eventual flight into space be designed to fly on the shuttle. If the shuttle had met its promised flight rates and safety levels, this would not have been a problem, but, as history's shown, the shuttle did neither. In 1986, as the transition was well underway, *Challenger* exploded soon after launch. In the accident investigation report, also known as The Rogers Commission report, NASA was ordered to return to the use of expendable launch vehicles, and only use the shuttle where human intervention was necessary. This divide took place rather quickly, and a few years later the shuttle was only flying science missions and new missions to the Mir space station to ferry personnel and cargo to the Russian outpost. In 1998, construction began on the International Space Station, and the shuttle had somewhere to go regularly, carrying new segments and large cargo runs for the ISS. In February 2003, *Columbia* disintegrated over the skies of Texas, and the accident investigation board concluded that personnel and cargo need to be separated entirely. As much as some people would shout that this approach wouldn't work for things like space stations, it already has. America's space station of the '70s, Skylab, was launched in a single flight of a Saturn V booster. Uncrewed launchers launched all of the Russian space stations, and the Mir space station was assembled over time with each part launched by an uncrewed vehicle. Even the ISS, assembled mostly through shuttle flights, had its start as a Russian launch of an expendable rocket. Later, a second expendable carried the

Zvesda module, which docked with the station automatically and serves as the backbone of the station.

Presumably, the *Columbia* accident board recommendation means any future space station or other object constructed in space would have its parts launched by expendable vehicles, and crews launched to work on it or in it would fly via another means than the cargo. Current systems, such as the space shuttle, will not fulfill this need. Launching a crew on the space shuttle with an empty cargo bay would be a huge waste, and it's only scheduled to fly until station completion, so something else will be necessary to carry just a crew and some supplies. NASA had its sights set on the Orbital Space Plane, but in the reorganization following the new space policy, development of the OSP has stopped in favor of the new, modular, Crew Exploration Vehicle.

I propose moving expanding this people/cargo split further, taking into account a proposed mission called Alternate Access to Space, where cargo that is of a "bulk" nature, such as water, food, air, etc., should fly aboard one type of launcher, while one-of-a-kind items fly on board another. We've already discussed the one-of-a-kind myth that, even though it exists, is only due to our meager actions in space right now. A higher flight rate will decrease the grip that myth holds on the space industry, and move space travel as a whole towards our everyday experience base. For now, though, one-of-a-kind hardware is the way things are, and items that fall into that category require an extra measure of care.

The "bulk" class of cargo allows some experimentation in launch vehicle design. New vehicles can be used to deliver bulk cargo, as can upgraded versions of current launch vehicles, like the EELV or the shuttle stack. The key here is that the loss of an individual mission is not critical, because there's plenty of air, water, or food around to send up on the next trip. It's possible, however, that an extended streak of bulk-mission launch failures could cause a crisis situation on orbit, with a space station or other outpost running low on critical supplies. If this is the case, the next re-supply mission is not classified as a "bulk" mission, but given the full scrutiny of a one-of-a-kind hardware launch or a crewed mission. The experimentation here is pretty open-ended, allowing orbiting supply depots described earlier, as well as relatively cheap resupply flights to the ISS.

In time, as hardware becomes common and launch success rates rise dramatically, this distinction will not be necessary. We don't put extra scrutiny on aircraft depending on whether they're flying cargo or people, both have to comply with FAA regulations. Granted, there's a simplification there since all winged vehicles capable of carrying people or cargo are still piloted, and therefore have people on them. It's possible that, eventually, spaceflight could achieve the same stature, with cargo spacecraft derived from passenger craft or vice versa, each crewed by a team of courteous professionals dedicated to making your (or your hardware's) flight safe and comfortable. One can hope….

CHAPTER SEVEN

EXPANSION OUTWARD: WHERE DO WE GO FROM LOW EARTH ORBIT?

Here, we'll take a look at all the places I see humans as able to reach in the next 10-50 years. In the spirit of full disclosure, I think that Mars makes the most sense as a destination, with the hardware developed to travel to Mars used to explore other sites, either as tests for the eventual Mars mission (an example of this would be the moon), or as concurrent exploration destinations (such as the asteroids). Once such travel were proven, the individual strengths of the asteroids and the moon could be exploited for benefit, as we establish a foothold on the surface of Mars. Let's look at some other destinations as well, though.

HIGH EARTH ORBIT

I use the term high Earth orbit to describe a group of destinations for human spaceflight. In high Earth orbit, spacecraft can enter an orbit that takes it far away from the Earth, with a guaranteed return, or they can travel to any number of "stable" points between two bodies, such as the Earth/moon or Earth/sun, and stay there for a long period of time. The locations are called Lagrange Points, named for the mathematician who first predicted their existence, and there are 5 of them for each body pairing. The first one, L1, is the most obvious, and lies on a line between the two bodies. L2 is on the far side of the smaller body, while L3 is on the far side of the larger body. L4 and L5 rest in the orbit of the smaller body, leading it and trailing it. For convenience, let's choose to define high Earth orbit as any place where people can't get back to Earth within a day through most of the trip. This definition includes the Lagrange points (L1-L5 for both the Earth-moon and Earth Sun systems).

An important point to make here is: if traveling in low Earth orbit is simply going around in circles, adding little to human knowledge that isn't already there, what would be the advantage of going to high Earth orbit? Well, there are a couple things.

The Goal

The goals for crewed high Earth orbit missions are, for the near term, largely equipment checkout and initial crew radiation exposure experimentation. High Earth orbit locations have an advantage over low Earth orbit stations in that they create an environment very similar to what a spacecraft will experience in a long-term mission to another planet or an asteroid (see the appropriate section of this chapter). Some of these similarities include the radiation environment (see challenges, below), a near-constant source of solar power, and long-term communications capability with Earth. Unlike those long voyages, however, a crew that's in a high Earth orbit has the option of a rather quick return to the mother planet. Using one option, in the form of a highly elliptical orbit, the spacecraft will return to Earth automatically. If this type of auto-return isn't desired, a craft can be sent to one of the Lagrange points for a long-term stay, with a small rocket firing bringing the craft back in a matter of days or weeks.

One other thing that cannot be undersold is the view. Uncrewed spacecraft leaving the Earth/moon system have taken amazing images showing both orbs as crescents, and to have human eyes set upon this view for the first time will be an incredible moment. New technologies will allow others around the world to share in that vision with unprecedented detail. Now, this view can't be counted on for mission after mission of "Wow" from the public. While humans often show their tendency to get very excited about new things, they always get bored with something once new very quickly. Remember the Apollo program?

Although there are some missions that are done much better from high Earth orbit, these missions are best handled by uncrewed spacecraft. Some examples include on-orbit sentry duty and astronomy.

On-orbit sentry duty, where a spacecraft stays near the Lagrange point directly between the Earth and sun, monitoring our star for storms that impact

Earth and nearby space travelers. The Advanced Coronal Explorer, known as ACE, is currently carrying out this mission.

The James Webb Space Telescope (JWST), originally known as the Next Generation Space Telescope or NGST, will carry out astronomy, essentially looking at objects in space. The JWST will travel to the 2nd Lagrange point, away from the sun on the Earth-sun line, where it will stay and look at the universe in a new light. This location will give the telescope a constant temperature environment with constant power supplies, and should provide an excellent view for years.

Trying to force humans into these missions would be a bad idea. In the case of astronomy, any human on board messes up pictures simply by moving around. Sentinel duty is extremely dull, and when the spacecraft detects a solar storm headed for Earth, the last thing you want to worry about is having a person on board that spacecraft. The storms that you're trying to detect are dangerous!

This is not to say that high Earth orbits don't have future potential, just that the missions we're doing at those positions right now don't translate well into a crewed mission. I've heard of a proposed "return to the moon" plan that uses the Lagrange point between the Earth and the moon as a refueling station. This plan uses current launch technology to land people on the moon again. There are no physical laws that preclude such a mission.

Another way that Lagrange points can become more useful is by placing an asteroid there. This is the sort of fiction-soon-to-be-fact proposed by John Lewis in *Mining the Sky*, and is absolutely feasible, assuming that enough energy can be brought to bear on an asteroid. Once an asteroid is an a Lagrange point, then the material within that asteroid can be mined and shipped back to Earth rather easily.

Others propose[62] that building colonies in the High Earth Orbit spots, with the number of missions needed to build them, is the step necessary to bring down launch costs and enable all the other journeys described in this chapter. While at its core, this theory has merit, some of the additional points brought out in the argument do not hold up. In the production of these colonies, there is an assumed shift from government-run space launch to commercially run space launch. I assert that any truly massive space effort, on the order required to build a colony at L5, would not experience that shift because of the huge expenditure to lift so much mass by the methods we use today. There would have to be at least some motion in the low-cost-to-orbit arena before such a colony becomes feasible.

Zero-gravity manufacturing is often cited as a reason for colonies in space, and it's true that truly zero gravity is feasible freeflying in space while it's not feasible on a planet surface. The problem comes in that truly zero gravity will only take place at one point in a colony—at the exact center of mass of that colony. Unfortunately, the center of mass for a colony in space will always be shifting, as people, fluids, and equipment move around it. These shifts are imperceptible to the people working on some zero-g manufacturing, but the materials they're working on would definitely notice the change.

Using the same logic of a massive number of flights to High Earth Orbit to lower costs, some propose building massive solar arrays in space. These arrays would collect solar power and beam it to Earth to meet our growing energy needs. Solar arrays in space get more sunlight to fall on them because they're above Earth's atmosphere, and because they rarely fall under Earth's shadow, they produce power almost continuously. On the surface, this argument makes sense, and it appeals to anyone trying to create a pollution-free society. Again, there's a problem. When the costs of constructing, launching, and maintaining solar arrays large enough to meet any percentage of our power needs are calculated, including depreciation of the assets, the break-even line for the system lies at a launch cost per kg to orbit of about $4.30/kg, which is a cost reduction of 2000 times compared to current space flight. By most accounting, this number is impossible to reach.

The Challenges

Traveling to high Earth orbit on an equipment checkout mission is by no means an easy task. It is easier than many of the other destinations out there, but the high Earth orbits provide their own set of challenges. Overcoming those challenges will open the door for the next round of exploration.

Radiation

Here on the surface of Earth, we live in a relatively low-radiation environment. We do not live in a radiation-free environment, no matter what any anti-nuclear protestors say. Current scientific theory holds that radiation in one form or another lead to many of the mutations that happened throughout history and pre-history, leading to the world we live in today. When you compare Earth's surface to outer space, however, we live in a relatively low radiation zone.

Our planet has a multi-level defense against multiple types of radiation. Our atmosphere, relatively thick and made up of lots of different gasses in various forms, blocks rays in the form of ultra-violet light (not completely, or sunscreen wouldn't be required), x-rays, and some galactic cosmic rays. Travelers on board the space station travel above the atmosphere, and are exposed to a larger dose of these radiations than the average person. In fact, even people that fly on commercial aircraft often (think pilots and flight attendants) are exposed to a larger dose of radiation than most people who stay on the ground throughout the same timeframe.

While living in low Earth orbit is worse than living on the surface as far as radiation goes, we have a pretty good idea of how this level of radiation affects humans. Decades of near-continuous spaceflight between the US and Russian space programs have exposed people to this environment for long periods of time; some travelers have soaked in the low Earth orbit radiation environment for over a year with no demonstrated health effects.

The next layer of defense for the Earth is its magnetic field. If you've ever used a compass to navigate, you've worked with Earth's magnetic field. Navigation is not the only gift that our planetary dipole gives us, however. Because of the way charged particles react when they travel through a magnetic field, the field itself protects Earth from a lot of solar radiation and some galactic cosmic rays. The sun constantly sends charged particles (usually protons and electrons) at Earth, and these particles hit the magnetic field. When they do, the field makes them move in tight spirals within the field. They continue to follow this curved course until they run into something, usually our atmosphere near the poles. This effect is visible to the naked eye in the northern and southern latitudes and is known as the aurora. When the sun is more active, that is, shooting more particles at us, the aurora is more active and can be seen closer to the equator.

This magnetic field is a concern for space travelers for two reasons: 1. Spacecraft leaving Earth orbit have to travel through the magnetic field to get away from the planet, and 2. Spacecraft don't have a magnetic field, so they have to deal with the charged particles directly once they are outside of Earth's protection. We'll deal with the second concern in a couple moments. The first concern is actually one of the first things discovered about the space environment. Back in 1958, when the United States launched its first satellite (Explorer 1), the vehicle carried a charged particle detector on board. Explorer 1 found that there were bands of charged particles encircling the Earth, held in place by our magnetic field. These bands or "belts" were named the Van Allen radiation belts after the scientist who suggested putting the particle counter on board. The radiation levels in these belts are something to be concerned about, in that you wouldn't want to spend a lot of time in them, but they're not a blockade. Nine Apollo moon missions traveled through the Van Allen belts on their way to the moon, but they sped through them so quickly that the crew was not impacted at all. One Gemini mission actually lingered in the lower belt for a short period of time, setting a new altitude record when things like that were all the rage. After a couple orbits, they brought their spacecraft down to a more common altitude for the rest of their journey. So we take care of our first concern by flying the way we did 35 years ago…with speed and a little bit of radiation shielding, which our spacecraft has anyway.

So, our spacecraft has raced its way through the magnetic field, and now it's off to a comet. What's this environment like? It turns out that the sun and interplanetary space play a bit of a tug-o-war as far as which one gets the headlines for the biggest radiation issue.

The sun is the most obvious object in the solar system. The boiling mass of gasses that keeps everything in our little biosphere going produces a lot of nasty stuff as by products, mentioned above. The standard things that you build a spacecraft out of can block most of the nasty stuff, protecting a crew.

In many ways, the sun is stable, and it's a good thing that it is, or else we wouldn't be here. There are some changes that happen over time on the sun, and they change the Earth in ways that we don't fully understand yet. On a spacecraft, without the multi-layer defense of Earth, these impacts will be something more to worry about. The primary sun cycle that humans have studied so far rises and falls in an 11-year period. During "Solar Max" or the time when the sun is most active, huge clumps of dark masses called sunspots move across our star's surface. Solar flares, ribbons of gas the size of many Earths, arch into space and back down. Coronal mass ejections, remnants of great explosions on the surface of the sun, fly off into space, loaded with charged particles that can play havoc with a planet's magnetic field, or with a spaceship's crew on their way to some destination. CMEs are the biggest challenge the sun provides for spacecraft makers. During "Solar Minimum," usually five and a half years later, the same events happen, but with much less intensity and frequency.

The other form of radiation, coming from outside our solar system, is known as galactic cosmic radiation, or GCR. GCR is made up of rather heavy atomic nuclei (think heavier than helium, more along the lines of iron) moving at amazing speeds. Some particles move so fast and with so much energy that the Earth's magnetic field doesn't even affect them. They crash through the field, our atmosphere, and us all the time. These particles have the ability to do tremendous damage to electronics (whenever something really strange happens to a piece of household electronics, I blame it on GCR—it sounds reasonable, and who's going to prove it wrong?) and perhaps human DNA. Earth protects us somewhat, but crewmembers of spacecraft will be much more vulnerable.

The ironic thing is that, when the sun is at it Solar Max, the GCR threat is at its minimum. The theory goes that when the sun is at maximum, it is throwing so much more stuff out there that the GCR can't penetrate as much. So, we could choose to launch missions when we're at Solar Min, and deal with the GCR, or we could launch at Solar Max and deal with the sun. The problem comes in that making that type of decision drives the times that you can fly. If you want to fly around Solar Max, you can only fly for a couple years around that period, which comes every 11 years. So you have to deal with both. It turns out that the CME danger from the sun is pretty directional. If the sun throws a big cloud of gas and charged particles at you, you know that it's coming from the sun. This lets you build directional shields to protect the crew. For example, the sun-facing side of the craft could store the mission's water supply. Water is an excellent shield, and therefore it would serve two purposes. By the way, the charged particles that the sun throws are not powerful enough to make the water radioactive.

GCR is a little tougher to deal with. Because of their high energy, the particles will race through just about anything that could possibly be put in their way, except for tons of dirt or an entire planet. Many of the rays would simply pass through the craft, unaltered by the experience. When a particle of GCR strikes a molecule or atom within the spacecraft, however, it does so with such energy that a "shower" of particles and other forms of radiation is formed. In this case, any shielding you used to keep the radiation out will work to keep it in.

How to, and whether we need to, defend against GCR may be the most pressing question in human travel beyond Earth. While some maintain that the GCR threat is minimal, others claim that the elevated level of GCR exposure, mixed with the particle showers that it will cause, can be debilitating for a crew. So far, thanks to the Apollo missions, we have proof that a small sample size of white males can function for periods of up to 12 days in the interplanetary environment. The only direct effect of GCR so far is that most of the Apollo travelers reported seeing flashes of light while their eyes were closed on their trips, likely caused by GCR. The flashes were distracting, but not debilitating. Long-term studies of health side effects on these men are ongoing, though old age is becoming an increasing factor for our first emissaries to the moon.

On a side note, the radiation environment away from Earth is one of the "facts" cited by moon conspiracy people as making the case that we couldn't have gone to the moon. They've even been quoted as saying that a human can't survive in the environment outside of the Earth's magnetic field. What they fail to mention is that the human they're discussing would not be in a spacecraft or a space suit. Well, there are a lot of reasons that an UNPROTECTED human can't survive outside of the magnetic field—lack of air to breathe, lack of pressure, extreme temperatures, and radiation. That's why we put people into spacecraft when they travel. Pretty smart, eh? Keep that one in your hip pocket for sometime someone asks you about whether or not we went to the moon. If you're looking for a more complete dissertation on the moon-hoax theories, check the work of Phil Plait[63].

Long-Term Flight Without Resupply

While the long-term occupation of the International Space Station is an achievement, it is largely a repeat of the long-term occupation of the Mir Station by the Russians. Arguments about capabilities aside, both Mir and ISS are the modern-day equivalent of coast-hugging oceangoing vessels, always traveling in sight of shore. Yes, it's possible for the ship to put on lots of miles traveling up and back, but there's no new territory being covered. The oceangoing craft of old had an advantage over our atmosphere-huggers, in that they were carrying goods from one port to another, so people could see immediate benefits from the effort. Last I heard, federal regulations (or at least bureaucratic stodginess) prevented that type of trade going on involving the ISS.

The biggest crutch that the Mir relied on, and the ISS uses today, is the constant resupply from Earth. The space shuttle carries up large pieces of the station, but also supplies including water and big grocery store runs of food. Russian *Progress* cargo craft, although they're much smaller, carry crew goods, special foods, and props for commercials and Hollywood Oscar openings. They dock with the space station several times a year. In times of crisis, like right now with the space shuttle grounded, that flight rate increases. Eventually, Europe will contribute to space station resupply with its Automated Transfer Vehicle. These trips are done without a crew—the *Progress* craft dock with the station automatically, and are marvels in their regularity, but the fact that we use them is not teaching us how to live beyond the cradle of Earth.

When we send a group of people on a shakedown mission into high Earth orbit, there will be no resupply. They have to have what they need with them when they leave. This has been done before, again in the Apollo program, and the Skylab program had crews on board for 84 days without a visit from another craft.

Psychological Effects

In my opinion, this concern is not nearly as big a deal as the other two, but enough effort is being put into researching it that I felt it needed to be mentioned. Psychological effects can be filed under two topics: crew dynamics, essentially how a crew will work together for long periods of time, and separation anxiety, or how a crew will function knowing that they're very far away from home.

I've seen some talks on the first topic, and get the impression that that work is focused on finding the "magic test" that will, when administered to team members, weed out those that won't work well when the pressure's on. The space travel rumor mill is rife with stories of simulations in Russia gone bad due to alcohol, hurt feelings onboard Mir, and bad experiences during grand explorations of the past. In my opinion, this is something that needs to be monitored, but can be taken care of with some factors like crew self-selection. This novel type of crew determination involves choosing a well-qualified group of people, and putting them into training together. They are told how many people will be on a team, and what requirements are necessary, and then they are put into a rotation of constantly switching training partners. After a time, the natural leaders will pull together teams of people who meet the requirements and can work together. Once teams are formed, they start training together for serious mission operations, and the best team is chosen for the mission in question.

I think that a major player in this group of concerns is public perception and mass media. Mass media works to draw in readers (or watchers) by appealing to things that people want to see, and the vast majority of the market prefers to hear bad (or even better, just controversial) news. Yes, it's true that a small craft designed to hold four people for 6 months will not provide the luxury that most people with the money to watch television enjoy. This "news scoop" doesn't even compare to the living conditions that early explorers faced, such as Nansen and Johansen, who spent 9 months in a shelter just larger than they were, waiting for the winter snows to melt in Greenland[64]. What if there'd been a hidden camera there for some sort of exclusive on a news show? The colonists who made their way across the Atlantic to start a new life in the new world traveled in deplorable conditions, and the explorers who went before them to map the new world lived in worse conditions. Why should people care today? If the people on board that craft are volunteers, and the people watching TV are not forced to live in these "horrible" conditions, why should it matter?

On the negative side of this issue is one incident that's drawn fairly wide attention in space circles, when a group of people serving in a Russian isolation lab (think of it as a space ship that's still sitting on Earth) ran into some problems during their isolation time. Reports are rather sketchy, but apparently during the holiday season a few members of the team enjoyed some vodka, and got a little amorous with other members of the crew who were not interested. Other members tried to intervene, and a fight broke out. The more sensationalistic versions of this story describe attempted rape and "blood on the walls"[65]. This is a serious event, and shouldn't be taken lightly, but I also believe that comparisons between this event and something that happens in space are overdrawn. Let's compare the two. In the first case, you are sitting in a can on Earth, missing holidays with your family, racking up exposure time for an experiment with other people who'll volunteer for the same treatment, only to walk out of the same door you walk in with perhaps a paper to write. In the alternative, real scenario, you are on the first ship going to Mars, with friends or at least peers whom you've trained with for years, and known each other longer, living in close quarters but moving ever closer to brand new land to explore as the pinnacle of a career in biology, geology, or chemistry. Here, I see two very different situations.

On the other extreme, I saw an astronaut speak at a Mars Society convention who said (this is paraphrased), "Put me in a 55-gallon drum with a proper supply inputs and waste outputs, give me a three-year supply of power bars, and let's go to Mars." Afterward, he added that the folks in psych hated when he said things like that.

Of course, the answer is somewhere in between, but I have to believe that it lies more toward the 55-gallon-drum version than the years of psychological research version. Think of how much unreported human history has taken place, a lot of which included hardship, before we even had terms like psychological factors, then imagine early humans not moving out of Africa because one of them came back a little scared after encountering ice for the first time. Let's face our fears and move on.

Advantages

High Earth orbit missions have some advantages that make them an excellent first step beyond low Earth orbit.

Constant Contact With Earth

Some other destinations beyond low Earth orbit create times when a crew will be out of communications with Earth for a period of time. While this difficulty can be overcome with planning and hardware, that work is unnecessary when traveling to high Earth orbit. Communications with Earth comes in handy, in case a crew falls on hard times and they don't have the answer to a problem. The people on Earth will be able to help them. Of all the high orbits and Lagrange points, very few put Earth out of communications for any time period.

Near-Constant Power

In our space travel so far, spacecraft have to take into account that they spend some significant portion of their time in shadow. The space shuttle does not rely on the sun for its power, so shadow times are not a problem. The ISS has huge solar arrays to gather the sun's energy, but must store much of it for times when the station is in the dark. During that time, the craft runs off batteries on board. The time spent in shadow is very near half of every orbit, so the batteries have to be very large. Travelers in a high Earth orbit will not have this concern. It actually takes a lot of work to be in the right place at the wrong time to put a high Earth orbit spacecraft into shadow. Of course, some batteries will be required for problems that may arise, and any craft that's being checked out for future travel beyond high Earth orbit will need its own set of batteries as well, but they will not be used too much on these trips.

Relative Ease to Return to Earth

This is where high Earth orbits for checkout really shine. Some Lagrange points, such as the one between the Earth and the moon, are about 2-3 days' travel away. If a crew is there, trying out some new spacecraft, and a serious problem develops, they only need to fire small rockets to send them on their way, and then stay alive for the short trip back to Earth. This Earth-Moon Lagrange point is the most flexible because of this. An actual high orbit of the Earth makes it a little more difficult to change return schedules, because once a spacecraft stops firing its rockets, the craft stays in that orbit until those rockets fire again. So, if a craft is to go on a four-month systems checkout, the craft will definitely come home, but it will come home in four months unless rockets change that. Because of that fact, I believe that the Lagrange points will win as far as high Earth orbit destinations.

Disadvantages

All is not rosy when discussing high Earth orbit missions, though. As time wears on, these disadvantages will grow.

Exposure Instead of Exploration

The fact is that high Earth orbit operations really don't take us anywhere. Of course, philosophically, it's not the destination that matters but the journey, but that sort of argument wears thin pretty quickly when you're either pressing for public funds or requesting more funds for a commercial project. Until something is there that requires work or upgrade (a space telescope), moved there (an asteroid), or built (a space station of some sort) in a high Earth orbit location that requires tending, mining, or habitation, their primary role will remain being good places to check out new equipment.

Some have proposed a Lagrange point as a stopping point for refueling before traveling onward through the solar system. This approach has its limited uses, as described earlier where the proposal was to send people back to the moon using today's hardware, but is not a good answer overall. Especially with missions to the asteroids and Mars, the effort it takes to get to a Lagrange point for a stop is quite similar to what it would take to race right past it and travel directly to your destination. Some proposals include building advanced-propulsion spacecraft at the Lagrange points, then using the new craft to fly beyond, but this actually makes trips longer than direct travel.

THE MOON

The Moon is Earth's nearest neighbor in the cosmos. It is incredibly obvious in the sky at night or in the day, and it's already been scanned from orbit by a series of spacecraft, with a small portion of its surface explored by robotic craft or humans in person.

The Goal

Goals for the moon vary. Some people and advocacy groups maintain that the moon is the natural choice for our first colony away from Earth. Others say that it's an excellent mining site, a place for small groups of workers to mine and refine rocket fuel, oxygen, water, and building materials for vast construction projects in the Earth-Moon environment. Still others call the Moon a siren, useful for some scientific study, but essentially barren and distracting us from more important goals beyond. We'll take a look at all three arguments.

For those who imagine a home-away-from home with unbeatable view, the moon is high on the list for future colonization. Although it's far enough for those who want to get away from Earth, from its surface the mother planet is still only a couple days away by spacecraft. If you just need to get a message back, the timeline for questions and answers is fast enough that a conversation is still possible. Our samples gathered from the surface show that there are lots of raw materials on our nearest neighbor, including metals, silicon, and oxygen. Data coming in since the early '90s show that there is also water-ice in permanently shadowed spots on the South Pole of the moon, but research in 2003 questions how easy that ice will be to access. The moon also provides a wonderful opportunity to stretch our spacefaring legs, toddling a bit in the relatively near-Earth environment before heading deeper into space for the asteroids or Mars.

While not excluded in the colonization plan, there are others who see the moon as a more Spartan mining town instead of a bustling colony. In this scenario, a group of roughneck entrepreneurs set up a mining camp on the surface, and perhaps some manufacturing to process raw materials into useful products. Once manufactured or simply purified, these materials would be shipped off the surface, either back to the home planet, or to another position in the Earth-moon system for use in building larger projects.

While the idea of shipping things off the surface of the moon back to Earth or into Earth orbit may sound a little strange, it actually does work as long as you're willing to make a fairly large assumption. First, you have to assume that there are people and equipment on the surface. These people, once there with the proper equipment (such as a mass driver, which is a very efficient, electrical way to get materials off the surface of the moon) with an infrastructure to keep them going, are an excellent source for materials. Starting from scratch, however, as we are right now with no people or equipment on the moon, the cost of "setting up shop" is significant.

So what resources could a moon mining operation provide? Helium-3 is the most likely and valuable answer, since the material is constantly put down on the surface of the moon in small, although measurable quantities. Extracting the helium-3 is not easy, involving scooping up large quantities of lunar regolith (dirt) and baking it to capture the precious gas, but there's a problem. Helium-3 is an excellent fuel for fusion reactors, and fusion reactors are kind of scarce here on Earth right now. In fact, current efforts towards developing fusion power are largely mired in an international commission trying to decide where the next developmental fusion reactor will be built. Until fusion power comes online, there will be no helium-3 market. When one develops, I'm sure that the moon will be a player. More on helium-3 in a bit.

Other materials, such as silicon, iron, etc. important to an industrial society are also found on the moon, but we haven't discovered a concentrated source of the materials as of yet. On Earth, volcanoes and other geologic events have pushed materials into clumps, which makes for excellent mining. Such mines are centers of activity for getting the material out of Earth's crust until the cost of drawing the material out exceeds the sale price of that material. In the case of gold, there's a mine in Victor, Colorado, that's on the verge of shutting down because of the relatively low price for gold on the world market. When the price of gold rises again, activity will pick up at the mine. For lunar mining, the cost of extracting materials is high, but the price of getting the workers and equipment there dwarfs the extraction costs.

One thing that a properly powered lunar base shouldn't have to worry about is oxygen. There is oxygen on the moon, locked in soil in the form of some metallic oxides. These oxides will take a lot of electrical power in order to break the bonds between the atoms and free the oxygen, but the process is relatively straightforward.

To me, even though it has its problems, the mining situation is a little easier to imagine than the colony condition described earlier. Since there aren't huge numbers of people to feed, it's possible to imagine that the miners can live off of food shipped to them from Earth instead of having to grow their own, which presents problems that will be covered shortly. Since this facility will be staffed by likely very well-paid volunteers on a relatively short rotation, the radiation concerns are less of a problem.

To me, the science available on the moon hits a homerun. The moon is prime real estate for observing the cosmos, both in visible light and radio waves. Many argue that such missions are better done in space, and there may be some special applications where this is true, but the moon offers its own ideals in science. First off, the moon carries a record of the Earth-moon environment, likely since very early in our solar system's history. Earth's surface is constantly being molded and changed by water, air, and volcanic flow, to the point where the surface can become unrecognizable in merely a few million years. The moon has been largely constant throughout that time, picking up more craters as the millennia march on, but always looking down with the same face on its mother planet. Since the surface hasn't changed much, it's likely that there's some pretty good stuff to find just laying around up there.

Ideas here vary, from ancient Earth rocks and rocks from the rest of the solar system through some more exotic claims of dinosaur fossils. The rationale on all parts is that basically any bit of cosmic debris that been floating around throughout the years in the Earth-moon vicinity would have landed eventually on Earth (and been incinerated in the atmosphere) or on the moon where it would land and be preserved. I find the concept intriguing, but I think you'll need to have a pretty fine-toothed comb to find anything interesting. The fact is that the same energy that will incinerate a rock or fossil that's approaching Earth will turn it into a gas almost instantly in an impact with the moon. I still think we should go look, though. There are some areas that are interesting for specific reasons as well, not relying on the possibility of tripping over some major find deposited randomly in an asteroid strike.

On Earth, structures like the Grand Canyon are important to geologists because the river that formed them cut through layers of rocks deposited over the ages. Because of this erosion, someone looking at the different colors of rocks in the Grand Canyon is literally looking back in time. While water doesn't flow on Mars today, and hasn't in a long time, there are other factors that can create a similar erosive effect. The moon has no water, and relies on impact craters for erosion. It turns out that the south pole of the moon is home to the largest crater on the body. Called the South Pole-Aitken Basin, the bottom of the crater is 12 kilometers (39,000 feet! Remember that Mount Everest is 29,028 feet tall) below the rim. Here, scientists speculate that the walls hold the same kind of significance for the moon that the Grand Canyon has for Earth. Though our previous research answered many questions about the moon, there's always more to learn, and it's likely that the South Pole-Aitken Basin will give us many of the answers.

Lastly, the moon will serve as an excellent base for astronomy. The darkness of a lunar night (all 14 Earth-days of it) will have tremendous value for visible astronomy. Since the moon doesn't have an atmosphere, visible light telescopes will also be able to do some serious work during the lunar day, as long as they don't point in the direction of the sun. We'll talk about this more in a bit.

The Challenges

Nuclear Power

The moon has one of the most difficult day/night cycles that humans will face in the near future. Since the body is tidally locked to the Earth (it turns at the same rate that it orbits, so you always see one side of the moon from here), the moon has a day/night cycle of 28 Earth days. For 14 days, at any one spot, the sun shines relentlessly on the surface, bathing the landscape in heat, light, and radiation. Temperatures on that landscape soar to over 125 degrees C. Then, things change radically. When the sun sets on the moon (a very sudden event—the lights are on one moment, then off the next because there's no atmosphere to create the beautiful sunsets that we're used to seeing on Earth) temperatures plunge, and on the far side, without the shining Earth above, the darkness is beyond anything humans can imagine.

For any sort of activity on the moon, the best source of power is a nuclear reactor. Yes, it's true that the previous lunar landings took place using only batteries on board the lunar module. In that case, the longest surface stay was 3 days, and the crew was in sunlight the entire time. For such a short period of time, batteries are the best option. Instruments that were left behind to study the moon in missions beyond *Apollo* 11 had small nuclear power sources with them, and those power packs kept the experiments functioning for years.

There are some that will argue for using massive solar arrays on the moon. Those arrays will charge batteries, or store their energy another way for use when the sun sets. The problem comes in that it's with the sun down that the base will require the most energy. When the sun is down, the living quarters will need to be heated, any outdoor activity will require a lot of lighting, and any plants being grown for food will require a separate set of lights. This system gets unmanageable quickly. First, the base would need a set of solar arrays that produce enough power to run the base AND enough power to charge batteries at the same time during the lunar day. Then, the batteries (or other storage device) have to be large enough to hold the charge (at the higher nighttime usage rate) for the 14-day lunar night. Other options, such as beaming power from Earth, are possible, but carry their own complications.

A nuclear power source, very low maintenance and always giving power, is the best option for this. Given the fact that nuclear power will also be important for the exploration of Mars and beyond, it is the only feasible power source currently available for these tasks.

Advantages

Always in Sight of Earth

Any base or outpost that establishes itself on the near side of the moon will have the same advantage discussed in the high Earth orbit section...constant contact with Earth. The moon has kept its same face towards Earth for billions of years, and it's unlikely to change anytime soon. What's even better, from the lunar surface's point of view, the Earth stays in the same place in the sky all the time. So, any antenna required to contact Earth doesn't need to move. It can stay pointed in the same direction, and requires the same amount of maintenance as the small satellite TV dishes popping up around neighborhoods around the world. After all, if something that small and simple can pass muster within a homeowner's association, building one for a lunar base should be easy, right?

Short Transmission Time

I had to list this as an advantage, although I'm not completely convinced that it is one. Anyone on the moon, being only 400,000 kilometers away, requires just about half a second to receive a radio signal from Earth. This makes two-way conversation possible, although probably a little difficult, much like talking to someone overseas on the telephone. If you've never experienced it, there is a little pause between when you finish talking and the person on the other end hears the last part of your statement. It takes a little getting used to, with a lot of people talking over each other when they first try it, but with practice, it can be done.

So, while it's true that this is "good," I have trouble calling it an advantage because it doesn't prepare future space operators for a time when communications like this will be impossible. Missions to all but the nearest near-Earth asteroids will involve minutes of delay for communications one-way. Mars has times when radio signals take 20 minutes to travel from the Red Planet to Earth. It takes twice that time to get a response.

Once communication times become large, the whole idea of "Mission Control" for space missions will have to be re-thought. Currently, if a crew is orbiting Earth, doing some sort of repair, they have to ask permission to turn a valve or power on some system while people on the ground verify that the crew is doing the right thing. How does this take place when there's a five-minute communications delay? What about a 20-minute delay? Imagine the load-up times surfing the internet....

Quick Returns Possible

As several trips already taken have proven, the moon is only three days away. If there's any sort of medical emergency or other reason that people need to leave the moon for Earth, they're a fairly quick shuttle ride away. Some studies have even been done showing that a 24-hour commute to the moon should be feasible[66], once the moon occupants start producing rocket fuel on their own.

Water (Likely)

The Apollo missions returned samples from the moon, and the rocks contained no water. Mainstream scientific thought held that there was no water on the moon for quite some time. Then, in 1994, a spacecraft called *Clementine*, a missile defense test cleverly disguised as a lunar/asteroid science mission, flew to the moon and orbited for a time, taking images of the lunar surface and using its own transmitter to paint the surface of the moon, *Clementine*'s experiments hinted that there was water in permanently shaded portions of the lunar south pole. In 1998, another mission named Lunar Prospector flew to the moon and painted a much better picture. Using on-board instruments, Lunar Prospector measured the neutrons that the lunar surface radiated while being bathed constantly with cosmic rays. Prospector discovered that there was a portion of the moon, around the south pole, that showed a decrease in radiated neutrons. To most scientists, the best cause for this drop in neutrons is a large deposit of hydrogen below the surface of the moon. It is thought that the most likely way for hydrogen to exist on the moon is as part of water ice. While I used a lot of words like "best cause" and "it is thought," the evidence is actually pretty strong that there is a large amount of water on the moon. I used those words to show how scientists draw conclusions based on links of evidence. For the record, the same deduction method was used to determine that there's subsurface water on Mars.

Helium-3

In the future, the moon may be the home of a new "gold rush." The goal in this rush would be to mine Helium-3, the fuel for the cleanest fusion power plants theoretically known of today. Here, I say theoretically because humans are unable, right now, to use fusion power to produce energy. We know that fusion power works in nature. If you want proof of it, look up in the sky on a sunny day. The sun uses fusion to give our planet and the rest of the solar system light and heat. We humans understand the process that goes on in the sun, and can create it for short periods of time in laboratories, but right now it takes more energy to contain our fusion reactions than we get out of the reaction. This would be similar to trying to stay in business when expenses are $10, and profits are less—it just doesn't work.

In time, it's likely that we'll work out the problems and start relying on fusion power for our everyday electricity needs. Once that happens, life on Earth has the potential to be transformed. Fusion power is safer than our currently used fission power, because it can be switched off quickly, and it produces very little radioactive waste. Fusion power will allow the creation of a much-hyped hydrogen-based economy, where hydrogen is produced from the cracking of water molecules, not requiring fossil fuels like today's process of chipping hydrogen off of methane molecules does, but the fusion process will require fuel.

There are several types of fusion reactions, and each has its advantages and disadvantages. The easiest to get "burning" involves the combination of the two heavy isotopes of hydrogen, deuterium and tritium. This reaction has the side effect of producing a lot of high-energy neutrons. These neutrons can destroy containment vessels very quickly, and, unfortunately, we don't have a system that can deflect them very well. Most fusion plans call for a magnetic field to hold the extremely hot reaction away from any metals that would melt when exposed to it, but neutrons will pass straight through that magnetic field. A reaction that is less damaging to its containment vessel is the Helium-3 and deuterium reaction[67]. The problem here is that He-3 is extremely rare and expensive on Earth. There's some on the moon, though....

The moon's surface is constantly exposed to a stream of particles from the sun. Apollo astronauts sampled this particle stream while they were exploring the surface, and analysis showed that this shower includes He-3 in larger quantities than found on Earth. Using this information, Zubrin[68] shows that a lunar mining operation is feasible that could produce helium 3 at a profit for Earth-bound fusion reactors.

Far Side's Radio Silence

If you want to go somewhere in the solar system away from the clatter of Earth, I can only recommend one spot for you. The far side of the moon is the quietest spot in the solar system as far as Earth radio frequencies go. All of Earth's transmissions—the radar, the space tracking signals, and worst of all the TV commercials, are gone. There are a couple sources of radio waves that will be visible, namely the spacecraft that humans have sent out to explore distant planets, but these radio sources are well recorded and can be ignored as required.

Why is radio silence such a big deal? Well, while many people are familiar with the kind of astronomy where an old scientist sits on a cold night and peers through an eyepiece into the grand vista of space, the most interesting astronomy is done without our eyes. It turns out that our eyes see only a very narrow part of the energy that's out there throughout the cosmos. An illustration, even though it's not completely correct, is to imagine wearing a pair of blue-colored glasses all the time. You'd see the world, but everything would be described in shades of blue.

Many events that happen in outer space give off radiation in all forms, using our blue-colored-glasses illustration, this translates to all colors of the rainbow. If astronomers and scientists limit themselves to looking at visible light, they'd miss most of what's happening, like us wearing our blue glasses. Astronomers who look across a broad spectrum of radiation are called radio astronomers. Ultraviolet, infrared, x-rays, gamma rays, microwaves, all these types of radiation are falling on Earth all the time. If you want proof, just tune your television set to a channel you don't get. Granted, this can be kind of tricky with cable TVs, but it was quite common a long time ago...in the 1980s. The fuzz that you see is your television picking up radiation from space, plus some other "junk" signals from here on Earth. Those junk signals are exactly what make a radio astronomer's life difficult. If you were on the far side of the moon, and took an old TV with you, the signal you'd see on your TV would be about the purest possible signal from space. Earth's drowning noise would be silenced.

So, if a radio telescope could be placed on the far side of the moon, with its own power source and a way to relay its data back to Earth, the impact would be incredible.

Why is the moon better than an orbiting station like the Hubble Space Telescope or the Chandra X-ray observatory? Because the moon is relatively stable. There are no moonquakes, and the sphere rotates quite serenely, keeping the same face towards Earth in its tidally locked dance. Any telescope that sets down on the moon will stay there...for a long time. Any other telescope that's launched to the moon to help the first telescope can land some distance away, and they will stay that distance apart...for a long time. By having multiple telescopes near each other looking at the same object, computer enhancement allows those telescopes to act as one huge 'scope, with its size measured by the separation distance of the smaller telescopes. Some have proposed this type of operation (called synthetic aperture) for a cluster of satellites in space, and it's true that the idea will work, but in my opinion it's much more complicated to keep three or more spacecraft near each other while simultaneously carrying out astronomy than it is to land a telescope on the moon, measure its position once, and start taking measurements. Landing multiple objects like this on the moon would make the practice relatively routine, and might lead to something else.

Here, I think that human-tending operations need to be looked at very closely. If there's an array of telescopes on the far side of the moon, the last thing we want to happen is to have someone show up in a rover and kick up a bunch of dust onto our mirrors. Eventual mechanical breakdown of the equipment will happen, though, and the risks of dust-coating optics will have to be weighed vs. the degraded performance of the system. That ought to lead to some great, heated academic discussions at the lunar astronomy institute sometime in the future! I can just picture the astronomers and engineers divided into different camps, one pro-repair the other pro-pure optics (it's bad to have your side described as anti-anything). There's a little insight into how I think.

I have to acknowledge one simplification here. There is at least one other spot that I can think of that shares this "dead zone" of Earth chatter. That spot is on the far side of the sun from Earth. Interestingly, there's a Lagrange point there (it's called L3), but there isn't a lot as far as real estate. If you want to settle down away from the chatter, the far side of the moon is the only place to go.

Disadvantages

Fuel Required to Land/Take Off

As near as the moon is (relative to other destinations), many people are surprised to hear that it takes more fuel to land 1 kg of mass on the moon than it does to land that same kilogram on Mars. This is because the moon has no atmosphere that can be used to slow down. A spacecraft that's going to the moon, following the flight profile that Apollo followed, has to burn its engines to leave Earth orbit, drift to the moon, burn its engines again to orbit the moon, then rev up the thrusters one more time and fire them the whole way down to the surface. Also, once a spacecraft is on the surface of the moon, it must burn an appreciable amount of propellants to return to Earth, weather or not it pauses in lunar orbit for a spell. In many cases, it takes more propellant to get back from the moon to Earth than to get back to the Earth from an asteroid. While there are proposals to install some form of mass driver on the moon, which could use electricity to accelerate materials (probably not people) away from the surface, a large amount of energy is still required. There's a simplification here, because a human mission to Mars would require a lot more food and other supplies aboard, meaning that the amount of fuel required would be the same or more than a lunar mission, but, for uncrewed craft and supply deliveries, the physics in a simple trade come down on the Mars mission side.

There are other ways to get to the moon. Some spacecraft (the Japanese *Hiten* craft, for example) fired their engines to get away from Earth. They did so at a time that made them "miss" the moon, at least at first glance. What these spacecraft did was "wait" at an altitude higher than the moon until Earth's natural satellite came by in its natural course. The craft then dropped a smaller satellite (named Hagoromo) into lunar orbit with a relatively small burn. This method of flying cuts down fuel use, but the tradeoff is time. The mission took 6 weeks to reach lunar orbit. This type of tradeoff will have to be examined before committing people to this type of travel, and even with the time trade, a lot of fuel has to be burned to land the craft on the lunar surface.

Day/Night Cycle

As mentioned before, in the section about nuclear power, the moon has a punishing day/night cycle. First comes 14 days of bright sunshine. Temperatures soar to 125 degrees Centigrade. This can be good, although you have to do a lot of work to protect yourself from long-term radiation exposure and the heat. Just when it seems that the sun won't end, the base is plunged into 14 days of darkness. Now the temperature drops to −125 degrees Centigrade. Granted, the Earth will be above, providing some light, but that won't be enough to keep plants alive in greenhouses, or provide heat to the to the base. Once again, this can likely be overcome with a constant source of power such as a nuclear reactor—it can make up for much of this radical change, and provide light to the inside of a base. The power levels required to maintain lighting on plants in a greenhouse, though, are very high.

Dust

Dust is going to be a problem in most places humans travel beyond Earth, and the moon has its own unique dust problems. Though it's not explosive in oxygen, as was suggested before the Apollo astronauts walked on its surface, the surface is covered with it and it will have to be dealt with.

The lunar dust is made up of what were rocks on the lunar surface. Over the eons, as asteroids pounded our natural satellite, the rocks were shattered into smaller and smaller pieces. That dust sat, absolutely still, until another nearby impact pushed it somewhere else or just shook it in place. This repeated shattering created lots of sharp edges on these dust particles, and the lack of wind or rain on the lunar surface kept those edges just as sharp as the day they were formed. So, when you see pictures of Apollo astronauts on the surface of the moon, covered in dust, think of the millions of tiny knives, spread throughout the dust, cutting their way through Buzz Aldrin's space suit. While no space suit worn on the moon could hold perfectly against the vacuum of space, even when they were new, the Apollo suits got progressively worse at holding air as they spent time on the surface. The lunar dust worked its way into suit fabric and into any mechanical connection, and wore it down.

Something will have to be done about the dust. There are a couple ideal options: 1. keep the dust off all equipment, and 2. make the equipment dust tolerant. I don't think that number 1 is possible, because of how much dust there is, and 2 is difficult, because anything that rubs together (think of wheels, suit fabric, things like that) can get destroyed quickly by exposure to this dust. So likely, lunar tenants will have to balance the two options. Build equipment as best they can, and accept the fact that some dust will get into it.

Lack of Carbon

Carbon is the basis of life on Earth. An entire science, organic chemistry, is devoted to how carbon forms unique materials, from life-giving DNA through energy-giving fossil fuels. Even though carbon is found almost everywhere on the surface of the Earth, the surface of the moon is almost completely barren of the atom. So there are no in-place sources for carbon on the moon. There will be sources of carbon that any people living on the moon will bring with them, but talking about it isn't too pleasant.

The fact is that the best source of carbon for use on the moon, the most efficient one, that can serve a dual role, is the food that travelers bring along with them. Of course, once the food's served its primary role as food, what remains is described in civilized circles as waste. For any sort of farming, plastic production, or other organic activity, this "waste" will be as good as gold. Now, there are methods that make the waste much easier to deal with.

For example, a process called super critical water oxidation[69] exposes such waste to high temperature, high-pressure water. This exposure turns the waste into its basic component chemicals. The carbon would end up as part of a soot on the inside of the processing container. Much easier to deal with!

Radiation

The moon's surface is exposed to a level of radiation almost exactly 1/2 the level of that a traveler would be exposed to in open space. The moon's bulk does an excellent job of blocking galactic cosmic rays (and solar radiation during the nighttime) from below, but there's nothing above our colonists, or miners, or scientific sentinel minders, to protect them from the GCR coming from the open space above them, or solar radiation when the sun is up at their location. All that shielding will have to be done through engineering, large piles of lunar soil, or toughening up our bodies to be able to handle the dosage. Another problem the radiation contributes to is the greenhouse problem mentioned earlier. Even if our station is desired to grow crops with some natural light from the sun (periodically shaded during the day and then lit artificially at night) the greenhouse will have to protect its plants from radiation as well. These difficulties are not impossible to overcome, but they do complicate life on the surface of our celestial neighbor.

Meteorites

The moon has no atmosphere, so the shooting stars that we see at night on Earth, along with their millions of invisible cousins, will head right down for the surface of the moon without stopping. This is not a show-stopper, because our spacecraft travel in low Earth orbit all the time, where the number of meteorites are approximately equal to which our moon-dwellers would see (remember, the moon blocks 1/2 of the sky when you're sitting on its surface, even more if you're near a mountain) and there are people aboard the ISS all the time. Luckily, the same type of shielding that will protect us from radiation (mainly dirt) will also protect us from meteorites in our pressurized homes on the moon. Working outside will be a different challenge, however, and any large windows will have to deal with the problem as well.

The (Near Earth) Asteroids

Theories about the development of our solar system hold that the sun burned at the center of a huge disk of dust, rock, and gas. Over time, in the inner solar system, pieces of dirt clumped together, growing larger and eventually drawing more dirt towards the expanding mass. This process formed the inner, rocky planets of Mercury, Venus, Earth, and Mars. In the outer solar system, the theories hold, there were more gasses than rock and dirt, so as portions of this mass gathered together, they formed the gas giants of Jupiter, Saturn, Uranus, and Neptune.

Between the rocky planets near us and the gas giants beyond Mars, something different happened. Either there wasn't enough mass to clump together and form a large planet, or a planet formed sometime in the past, but broke up over time due to Jupiter's gravity or some cosmic collision. Either way, what remains between Mars and Jupiter is a belt of planetary rubble known as the asteroids. Most of the asteroids we know of exist in this belt. Over time, some of the main belt asteroids had their orbits changed through close approaches to each other, Mars, or Jupiter. Some of these wandering asteroids cross Earth's orbit at times, and these are known as the near Earth asteroids, or near Earth objects (NEOs).

The Goal

Asteroids and other objects floating through the solar system likely played a major role in the development of our planet. Impacts from space are suspected of many things, from possibly delivering life-giving materials to a young Earth, through wiping out the dinosaurs and potentially altering the course of empires, as in the theorized impact that served as a sign in the conversion of Constantine[70] to Christianity. Understanding these objects lets us understand our past, and eventually protect and build our future. For the near term, we need to be concerned with NEOs.

The Challenge

There are lots of different kinds of asteroids near the Earth, where the different types are largely decided by the original size of the object when it was formed, and how much of it was chipped away by collisions with other asteroids through the years. Some asteroids, if they were large enough, formed a layered structure much like a planet. At the center, a metal core stayed molten for a period of time, then turned solid as the fires of creation died down in the solar system. Others, of a smaller size, stayed as loose clumps of rock, just waiting for the next object to cross its path—either another asteroid or a planet. If the approaching object is an asteroid, the two chunks of primordial solar system collide, breaking into pieces and exposing material that was deep down in each asteroid to space. If the object is a planet, the asteroid drops onto it, causing catastrophes such as the extinction of the dinosaurs on Earth (the jury's out on this one, it may have been a comet), or the fireworks show watched by the human race as the comet Shumaker-Levy 9 slammed into Jupiter in 1994.

There is a subtlety here that I've skipped. Comets, travelers from the outer solar system that make a periodic trip through the solar system to brighten our evening skies, are, like asteroids, leftovers from the creation of our solar system. They originate from the distant edges of the solar system, however, and therefore have lots of water and frozen gases on them. These materials are known as volatiles, because they change state from solid to gas rather easily. As a comet comes into the inner solar system, the volatiles on its surface, warmed by the sun, shoot into space. This gives a comet its trademark tail. Eventually, all the volatiles on a comet melt off (the term is imprecise, but serves the purpose here), and that comet becomes, essentially, an asteroid. Over time, this dead comet may pass by some of the major planets and change its orbit to make it a NEO. That makes it all the more interesting to look for!

Traveling to, and learning about, asteroids is important for three reasons. The first has to do with survival of the species, and the second two have to deal with building large structures in space for people to live on.

Let's deal with the important stuff like our survival first. This account is taken from Congressional testimony made by Brigadier General Simon P. Worden[71]. On June 6th, 2002, the Earth was undergoing a rather average day. World leaders were bickering about typical problems, brought into greater focus by the recent attacks within the United States, with perhaps one exception. Pakistan and India, the two newest confirmed nuclear powers, were in a tense showdown over a piece of land called Kashmir that both countries claim ownership of. The armies on both sides of a line were shooting at each other increasingly, and it looked as though the entire kettle might boil over into a regional conflict involving nuclear weapons.

In the middle of this, an event a few thousand miles away nearly impacted human history. An asteroid, never before detected, beat the odds and flew into Earth's atmosphere, over the Mediterranean Sea. This asteroid was not large enough to strike the surface of the planet, but instead exploded in the night sky, with a bang that could have been heard for miles, and likely would have broken windows on a ship, if one were directly below it. The flash was impressive, too. In fact, it could have been confused for a nuclear detonation.

The question comes in as…what would have happened if this asteroid entry happened over the land of Kashmir instead of over the Mediterranean? The United States is aware that this event was an asteroid, because of billions of dollars invested in sensing equipment and decades of experience using that equipment. The Indians and the Pakistanis have no such experience or hardware. Would a field commander in Kashmir (on either side) see that flash and say they were being attacked? Would a commander at headquarters wait for more information, verifying that this was a nuclear attack, or would they order an immediate counterattack? If that asteroid exploded over Kashmir, it's quite likely that the map of the world would look different today.

The asteroid experience described above is an exception to the rule, in that events like it don't always happen when the world is under such heightened tensions. There are several events like it recorded in recent history, though. The US military started paying close attention to them once people started realizing that asteroid impacts were happening all the time. Initially, the military's interest in these asteroid entries was based on the fact that the Air Force needed to be able to tell the difference between an asteroid entering Earth's atmosphere and an incoming missile. Once that work was done, Air Force interest in the phenomena dropped.

Perhaps the most widely known impact happened about 65 million years ago. The dinosaurs, long-time rulers of the planet, were facing hard times to begin with as Earth's continents shifted into their final forms. The numbers of giant reptiles were dwindling, likely cut down by the choking gasses spewed by volcanoes around the world. Then, as it is widely accepted in scientific circles, a massive asteroid or comet slammed into the planet, in the area now known as the Gulf of Mexico. This asteroid was the size of Mt. Everest—6 miles in diameter. The energy released as the space rock struck was equivalent to trillions of tons of TNT, and the explosion killed a significant group of dinosaurs on its own. At the same time, the blast threw huge clouds of dust into the air and blocked out most of the sun's rays reaching the Earth's surface. Over the next few months, 65% of life on Earth vanished.

Using telescopes and meticulously searching the skies, the human race is working to figure out where all of these "global catastrophe" asteroids are. Once we figure out where they are, and what direction they're traveling in, we'll be able to predict when one of them will threaten our planet, and when we need to have a defense ready. We'll talk more about that defense in a moment.

The exploding fireball over the Mediterranean and the dinosaur killer are the most common, and the largest events, respectively, but they are not the only ones by far. There are many possibilities between these two in asteroid size and destructive power. A report on asteroid impacts[72] describes the different types of threats, and small impactors stirring up trouble in world hotspots is one of the players where the asteroid itself is a minor threat. There are also examples where the asteroid impact itself was more of a threat than our Mediterranean example, but less of a threat than the dino killer. In 1908, an asteroid entered Earth's atmosphere and exploded over the surface, flattening thousands of acres of forestland in Siberia. Once again, this incident happened over an unpopulated area, so the human death toll was low. The actual number is impossible to count because the area was so lightly populated and under even less control. This "city killer" type of event is estimated to happen about once a century, which merely states the chances of the asteroid hitting the Earth. The chances of it destroying a city are significantly less because, no matter what the mayors or egos of some cities on this planet say or think, the majority of the Earth is made up of water or barren land. Of course, the idea of an ocean impact, with the accompanying tidal waves, does not sit well either.

"OK, OK," you're saying. "I know that there are asteroids out there, and that they've hit us in the past, and they'll do so again, but what do we do about it?" Well, that's part of the problem. We don't know.

Some people (mostly Hollywood types) say that we should blow the asteroid into a whole lot of pieces, just before it hits the Earth. Granted, this makes for at least a mediocre movie, but bad science. It is true that a nuclear device may blow apart some asteroids, but how do we know whether the incoming rock is the right kind to blow up? It turns out that an explosion like the ones depicted in Hollywood blockbusters are likely to be worse for our planet, not better. An explosion at the center of an asteroid keeps the center of mass of that asteroid in exactly the same place. While some will argue that it's possible to break the asteroid into small enough pieces to cause them all to break up in the atmosphere, it's more likely to turn a single huge impact into several very large ones, and no one can come forward and say that one is better than the other.

Other ideas, while they don't hold the visual appeal of the Hollywood solution, are more likely to work. The problem comes in that, since asteroids come in radically different types, each solution does not work for all the types.

Some scientists say we need to detonate nuclear weapons near the asteroid, but not right on it. According to them, the explosion (or series of explosions) will serve as a poor-man's rocket, and push the asteroid away from its cosmic encounter with our home. This solution may work for the more solid ones, but others that are likely made up of dust and loosely held boulders will respond to this in much the same way as the Hollywood solution—they'll break up and pepper our planet with rocks.

Other scientists say that, given enough warning time, simply painting an asteroid can make a difference. You see, the sun's light exerts a force on all objects in the solar system. Although this force is almost imperceptible, it can be harnessed to produce incredible speeds or previously thought impossible feats, such as holding a small spacecraft stationary in position between the Earth and the sun, outside of the Lagrange points. It turns out that the color of an object exposed to the sun affects the amount of force exerted. The theory goes that if an asteroid is scheduled to strike Earth sometime in the future, painting the asteroid will cause a slight change in the force that the sun exerts; over time this will change the asteroid's orbit enough to avoid our planet. This method appears to apply to both solid asteroids and loose conglomerates (though I imagine that the paint will have to be applied very gently in the latter case!)

The last possibility is the most likely to work, although it is the most difficult as well. If a spacecraft of some sort can land on the asteroid, and then, using its own form of propulsion, push the asteroid, the space rock's appointment with Earth can be delayed indefinitely. The design of this "tug" spacecraft comes under a large debate itself (amazing how that happens!) as some argue for a large nuclear rocket, some for fusion power, and others say that the mass of the asteroid itself (either volatiles used as fuel or pieces of rock pushed away) be used to provide the power.

The one thing that all three methods of "asteroid wrangling" have in common is uncertainty. While some seem applicable to all asteroids, it's difficult to say whether or not they'll actually work, because we don't know enough about the space rocks that we're dealing with. To date, spacecraft have flown past four asteroids/comets, and one spacecraft actually orbited an asteroid for a year. The data returned has increased our knowledge by an amazing amount. For instance, before the *Galileo* craft flew past the asteroid Ida in 1993, no one had proven that asteroids could have moons. Images returned by the craft showed a small rock, now named Dactyl, circling the larger. While other missions are planned to bring samples of asteroids to Earth or land on comets, some of their funding is in question.

The fact is, one or two landings on asteroids, while useful, will not answer all the questions we have about the other types of asteroids. Only by sampling each type, to the point where we know what holds them together and what could break them apart, can we have the full spectrum of information we need to determine what type of action can stop a space rock from hitting us. This massive sampling effort would need to be combined with ground-based work to look at asteroids with equipment we have here. Comparing what the spacecraft sampled to what our ground-based observations estimated would provide us with a realistic chance of being able to characterize rocks in the future.

There is an alternate plan that makes me a little nervous, but is likely to be more palatable to those who have trouble with long-term (for most, especially those holding political office, this is more than 2-6 years) planning. The human race can bet on a long lead-time. If an object is discovered that will hit Earth eventually, but not for years or decades, that object itself can be sampled. Using information from that sampling, a method to deflect the rock in question from us can be devised. Yes, this option provides the best information about the rock in question, but it relies on a bit of a leap of faith that there will be time to build a sampling craft, run that mission, then build and fly a diversionary mission. I know that there are options, like building several response mechanisms while awaiting the sampling results, but again we're relying on some significant warning. There's been more than one object discovered AFTER a close approach to Earth. Granted, these pieces were not "planet-killer" sized, but they could have caused events similar to the India-Pakistan event described earlier. Personally, I'd like to have a little more information to start, instead of relying on spotting a rock that'll strike us in the next century!

So, hopefully, I've convinced you that learning to move asteroids away from a crash course with our planet is a good thing. Once we master the art of deflecting asteroids, or perhaps as part of our apprenticeship in it, we may want to move one nearby. We'd do this at our own pace, and move the huge rock into place near our planet, likely one of the Lagrange points between Earth and the moon. Then we can use the asteroid for our own goals. This leads into the second and third uses of asteroids mentioned earlier. While these ideas don't have the splashy showmanship of being able to save the human race, they do feed into a future where people build things in space.

As described earlier, asteroids come in many types. We currently believe that many of the asteroids of today were once part of larger bodies. Through the eons, as collisions between the objects took place, large objects were turned into smaller objects. Based on pieces of asteroids that fell to Earth as meteorites, we have a feel for what different types of asteroids there are. Some are metallic, that is, they are thought to have formed at the center of a large asteroid, large enough to create a core similar to the one at the center of the Earth. After a time, that core cooled, and a huge impact may have split the asteroid into fragments. Pieces that came from the cooled core form the basis for metallics. These metallic meteorites will prove to be excellent sources of metals such as iron. Now iron is not an expensive material by today's standards, but think of how much it costs to make iron and launch it into space. Granted, in all likelihood, by the time our species is moving asteroids, the cost of launching such things will go down, but having a ready source of iron above the surface of the Earth will provide excellent resources to our growing space presence.

Other asteroids contain more precious metals. By sampling these space rocks in the form of the ones that fall to Earth all the time, we know that there are asteroids out there rich in gold, platinum, and other materials that would help make space manufacturing work. Also, once we've learned to move asteroids, it doesn't take much imagination to imagine our stopping an asteroid or a comet from spinning. Once that's happened, the permanently shady side of the object will become incredibly cold, and any volatiles (read as: water, ammonia, etc) will remain frozen. This rock would become another source of needed materials to support human life.

Science fiction writers have been fond of the idea of hollowing-out asteroids for decades, and there's nothing impossible about the practice. A solid, rocky asteroid could provide an excellent outer shell for a spacecraft, and it would already have a lot of radiation shielding on board. The humans inside could enjoy an environment comparable to Earth in its radiation levels. Granted, the inhabitants would likely want to put some windows on their rock spaceship, but when hollowing out asteroids is an option, putting windows in should not be a challenge.

Advantages

Relatively Short Missions

By definition, NEOs are near the Earth. There's at least one that almost qualifies as an additional moon of our planet, because it hovers so closely. On a practical level, these small distances translate to short one-way trips to the asteroids, making them attractive for a crewed mission in the near future. Some say a mission to Mars, with travel lasting at least six months in each direction, is a political non-starter because it's hard for people to imagine machines supporting people for that long. An asteroid prospecting mission, with a travel time of as little as 30 days, serves as a good middle ground between early missions to high Earth orbit or the moon and the longer missions. Throughout planning such missions, however, planners will have to remember that the faster the mission travels, the more propellants the craft must carry. More on this in the technical stuff.

Low Gravity

NEOs do not have a lot of gravity. Therefore, it takes very little fuel to slow down after entering their sphere of influence, or to move out of their sphere of influence. As proven by the NEAR spacecraft, which set down on Eros even though it wasn't designed with an asteroid landing in mind, landing on the surface is also not that difficult.

I worded that last sentence very carefully, because any prospectors approaching an asteroid from Earth will have to use rockets to cut their approach velocity (see the disadvantage "no atmosphere" listed below). Remember, that if you need to travel a certain distance, the higher speed you travel, the sooner you arrive. Well, when traveling in space, you have to slow down every bit as much as you speed up. (This is also a simplified statement, but it holds essentially true for this discussion. If you want more detail, see Chapter Eleven, "The Technical Stuff.")

This low gravity can make for a very flexible mission. A spacecraft doesn't need a lot of fuel for landing and liftoff, in fact it may be possible for some NEOs that a spacecraft wouldn't need to land on the body at all. Crews could simply drop from their craft to the surface of the asteroid to explore wherever they liked. When the time came to leave for Earth, a small rocket firing would take them away from the asteroid, followed by an additional burn to set them on a course for home. Of course, the desired trip length would set how big the second burn needed to be.

New Territory

The first landing on an asteroid will be as pivotal a moment in human history as the landing of a person on the moon. It's likely that this won't be immediately obvious to the public, largely because asteroids aren't something that everyone on Earth can look up and see most nights or days. While asteroids suffer from this initial public relations problem, when the first one is deflected from an impact with our home, stock in missions to them will go way up.

So, for the first time since 1969, the human race will be looking beyond the next mountain. While I find it difficult to express the importance of this in words, I don't think it can be overstated. The last lands on Earth have been explored. Human feet have already treaded upon the moon, even though it still holds many mysteries. The asteroids, with all their potential positive and negative impact on human history, are virgin territory. No one has stood upon an asteroid and looked back at our home, the world Carl Sagan described as a pale blue dot, reminding us how small and precious our planet is.

Minerals/Makeup

I've been calling the first travelers to an asteroid prospectors for a reason. Just as the original gold rushers headed out to Colorado, California, and eventually Alaska in search for riches, the asteroids hold similar promise. Granted, it will take a little more than a pickaxe (to dig) or a pan (to search streams) for these riches, but they are out there. As mentioned before, some asteroids contain the same iron makeup that's estimated to be at the Earth's core, while the estimates for others run off the scale in rare earth elements and precious metals. There's gold (and silver, and copper, and platinum) in them thar rocks!

A prospector on the surface of one of these rocks, with proper seismic gear, will be able to determine the structure of the asteroid in unprecedented detail. Core samples taken from below the surface will give an idea of the mineral makeup throughout. All of these tests will have to be run rather delicately, though. Because the wrong size seismic charge will push the equipment (and potentially the prospector) away from the surface, and may create a difficult situation that's discussed in the disadvantages next.

Disadvantages

Low Gravity

Even though low gravity was listed as an advantage of asteroid exploration, because the low attraction makes landing and leaving easy, in some cases it makes leaving too easy. Every object (planet, moon, asteroid, comet) in the solar system has a value attached to it known as escape velocity. Any object (or person) traveling as fast or faster than the escape velocity will leave the body in question, never to return. On Earth, where the escape velocity is measured in thousand of meters per second, this is not much of a problem. On a near Earth object, the escape velocity may be mere meters per second, and a human that can jump to a height of one meter on Earth achieves a velocity of 4.4 meters per second!

The escape velocity of an object varies depending on the size and mass of the object in question. Large, heavy objects such as planets have high escape velocities while light small objects have low escape velocities. This will have to be taken into account when we're choosing the asteroid we want to visit, but it won't be a show-stopper.

No Atmosphere

Planetary atmospheres come in handy a lot. Earth has a wonderful atmosphere that lets us breathe and also keeps the rest of what we need (mostly water) around us so we can live. Not all planets' atmospheres are this useful, but they can come in handy for other things. We'll talk more about what an atmosphere can do in the next section on Mars, but for now, the lack of an atmosphere weighs against visits to asteroids. Earlier, when we talked about low gravity of a NEO being an advantage, we mentioned that the rocket power to enter orbit around or land on an asteroid was pretty low because of its low gravity. It turns out that you'll still need to bring all your rocket fuel with you when you go, because there's nothing else to slow you down. An atmosphere would do nicely.

MARS

Mars is my favorite destination beyond low Earth orbit. To me, Mars represents the smartest destination because of its resources and atmosphere. Also, since it's the destination that's the farthest away in the inner solar system, equipment built to go to Mars enables trips to all the other locations discussed so far.

The Goal

Mars is many things to many people. In ancient times, its perceived red appearance earned it the name of the Roman god of war. Having blood-red Mars above the horizon during a conflict was perceived as a good omen for an upcoming battle. As time wore on, and astronomers turned their telescopes to the red planet, Mars held humans' hopes for life beyond Earth. The publications of Percival Lowell, describing a dying Martian civilization, held the public's interest for years, and can still be traced as the roots of some of our fascination with the planet. Such stories, amplified by the much-hyped Halloween prank that Orson Wells pulled on the United States in 1938, kept the idea of Martian life alive, even though the facts supporting it were extremely sparse.

The dawn of the space age and its missions to Mars dispelled most of the Lowell mythos. The first spacecraft to photograph Mars, swinging past the southern hemisphere, showed a planet that looked a lot like the moon, with craters and rough terrain reminiscent of our nearest neighbor in space. An experiment done as the craft passed behind the planet proved that Mars had an atmosphere, although it was much thinner (on the order of 1% as thick) as Earth's atmosphere. The tide of scientific opinion turned away from the dying lands of Lowell, toward an already dead land that likely never held life.

The next shift in Mars opinion came in 1971 when *Mariner* 9 became the first spacecraft to orbit the red planet. Because this spacecraft flew over most of the planet, it was able to take pictures of the northern and southern hemisphere, and pointed out an amazing difference between the two. While the southern hemisphere was jagged and crater-scarred, the northern hemisphere was extremely smooth. The smoothness of the northern hemisphere reminded many scientists of the contour of Earth's ocean floors, and planted the seeds of the ideas that perhaps Mars was wet at one time. Divisions between the smooth and rough patches were combed looking for traces of a possible shoreline, although the technology of the time made it difficult to spot such a thing in the *Mariner* pictures. *Mariner* 9 discovered the largest canyon in the solar system, which now bears its name as discoverer—Valles Marineris stretches 4000 km (2500 miles) across the surface of Mars.

In 1976, two spacecraft named *Viking* landed on Mars with the mission of determining, once and for all, whether there was life on the surface of Mars. Each craft was made up of an orbiter and a lander. The spacecraft entered Mars orbit, and then the orbiter took photographs of the Martian surface, scouting a landing location for the lander. The landers separated from their mother ships, then settled down to the surface of Mars on July 20th and September 3rd, respectively. Once on the surface, the *Viking* 1 and 2 landers set about proving the existence of life on Mars. Scientists outfitted these robots with numerous experiments that were thought to provide the ultimate answer to the ultimate question.

The first experiment used was called the labeled release experiment. This device assumed that any life on the surface of Mars was dormant because of the lack of water there. The *Viking* lander picked up a scoop of dirt and placed it in a reaction chamber. Then, a small amount of water and nutrients (think of fish food) was added. The nutrients were marked (or labeled) with radioactive carbon, the thought being that any "bugs" on the surface would eat the nutrients and produce carbon dioxide, like creatures do here on Earth. Any radiation levels in the gasses above the Martian soil were to be interpreted as life.

Both landers carried out this action, and much to everyone's surprise, the radiation detectors gave a reading that was consistent with life. As soon as the nutrient and water solution was added to the soil, radioactive carbon dioxide was read in the gasses above the sample. Scientists were elated that they'd proven life to exist on Mars, but Mars had another surprise waiting for them.

Another test cast these findings into doubt. The gas chromatograph system (GCMS) was designed to break the soil into its component chemicals, and look for organic compounds, the building blocks of life. At the time, it was believed that any life found would be carbon-based like life on Earth, and a lack of carbon in the soil would mean that there was no life on Mars. The GCMS ran its test, and the result came back that there was no carbon in the Martian soil.

The scientific community was stunned. How could the LR experiment provide such positive results for life and the GCMS provide such negative results for life? One theory was that the LR results came because of a reaction in the soil where chemicals produced the telltale gas instead of the "bugs." In this crestfallen state, the lead scientist of the Viking team made the statement "if there's no organics on Mars, there's no life on Mars. Case closed."

Since that press conference a lot has happened that pried the case for life on Mars open just a bit. First, we humans learned much more about life here on Earth than we knew in the '70s. From extremeophilic life that thrives in conditions such as high temperature and pressure, through other life forms that live within rocks themselves, including methanogenic life that eats methane as a primary nutrient, we now classify life using much clearer vision than we did thirty years ago.

The scientists who worked on Viking have moved in their own directions as well. One of them, Dr. Gilbert Levin, a primary scientist on the Labeled Release experiment, maintains that his experiment found life on Mars. In his argument, he states that the GCMS was not able to find organics in some Earth-based test samples where organics were known to be, so the fact that it found no organics in the Martian soil means that organics may still be there. He also notes a day/night cycle to the generation of the gas, insinuating a dormant period for the life forms. He's proposed follow-ons to the LR experiment that he believes would exclude chemical processes as a reason for gas formation. Through his work, he's found a sugar substitute that's enjoyed marketing success[73]. His idea to fly miniature, upgraded LR experiments on missions has not been approved, probably because of two things. First of all, the surface of Mars is deadly to every form of life we've found, even with our newly expanded understanding of life on Earth. The Martian surface is constantly bathed in ultraviolet light (an excellent life-killer, as evidenced by its use in water treatment plants to purify drinking water) and it has a number of peroxides mixed in the soil, which could explain the results of the LR experiment and are also quite efficient at wiping out life forms. There is probably another factor, which is the desire to prevent a build up in excitement in "Life on Mars" research followed by an ambiguous answer that most on the Viking program believed was found.

After Viking, space agencies turned their eye away from Mars for some time. In 1992, a spacecraft called Mars Observer launched, aiming to renew our scientific assault on the planet. Unfortunately, contact was lost with the mission just before it arrived in orbit. It's believed that the propulsion system, planned for use to put the spacecraft in orbit, exploded and destroyed the craft.

In 1996, as mentioned before, NASA created a sensation by announcing the potential for microfossils in a rock from Mars found on Earth. At that time, two missions were being put together to go, in the form of *Mars Pathfinder* and Mars Global Surveyor. *Mars Pathfinder* set the internet on fire, as record numbers of people signed on to see the latest pictures from Mars and to track the progress of the little Sojourner rover. Pathfinder got most of the popular press, but *Mars Global Surveyor* will likely go down as one of the best buys in space history. The craft was built for about $150M, and at this writing, is still operating in orbit around Mars. Some of the more spectacular finds it has presented are a complete relief map of Mars, giving us a feel for how strange a planet it is, as well as proof of recent fluid flow on the surface, which could be water.

SPACE: WHAT NOW?

A dark time followed for Mars exploration, as both an orbiter and a lander, scheduled for arrival in 1999, succumbed to various technical problems that may or may not have been caused by funding or management problems. The orbiter traveled the whole way to Mars, only to come too close to its goal right at the end and plunge too deeply into the Martian atmosphere—something it wasn't designed to do—and burn up. This missed target found its root cause in the rather simple conversion from the English system to the Metric system that school children have been taught for years. The technicians involved could have done the conversion easily, but their documents didn't specify that the conversion was necessary. The lander is perhaps even more frustrating, in that the most likely cause is the lack of an "all-up" systems test. Most spacecraft parts are meticulously tested as they are assembled, each part proving its worth usually above and beyond what is required, before they are put together into a whole spacecraft. Then, the spacecraft is tested, to the maximum extent possible, in all its flight conditions. For the Mars Polar Lander, this full-up test didn't happen. It's very likely that a full-up systems test would have uncovered a flaw in the software that prematurely shut down the engines and likely caused the crash. The budget is partially to blame, as testing is notoriously cut whenever the budget axe falls and cuts a project's funding. Another factor, not mentioned as much, is overconfidence. Having taken part in many space projects, I don't find it hard to imagine a group of mid-to senior-level managers convincing themselves that less testing is required than originally called for. Putting the event in context, The Mars Polar Lander mission was following the enormously successful Pathfinder mission, which definitely cut a few corners in systems, but still pulled off a spectacular popular and scientific event. In the case of MPL, at least one too many corners were cut.

After the loss of the *Mars Climate Orbiter* and Polar Lander, dramatic cuts took place in future Mars mission plans. Instead of launching both an orbiter and a lander to Mars at each opportunity, one type of spacecraft would fly. An orbiter, named 2001 *Mars Odyssey*, traveled to the red planet in 2001, carrying a heat-mapping camera, a device to measure radiation levels in Mars orbit, and a device designed to map Mars' surface for water and other minerals. This neutron spectrometer watches the surface of Mars, looking for neutrons. There's a normal number of neutrons per second that come from a dry surface, and when that number decreases, or the neutrons themselves become less energetic, the spacecraft notes its location over Mars. It is said that these slower or fewer neutrons are moderated, and one of the best neutron moderators is hydrogen. One of the most likely substances to contain hydrogen is water, and therefore, it is very likely that there is water, probably in the form of ice, within the first few meters of the surface at that spot.

In June and July of 2003, two Mars Exploration Rovers (MER-A, and MER-B, or, as named by Sofi Collins from Scottsdale, Arizona, *Spirit* and *Opportunity*) started their journey to Mars. These two identical rovers are designed to function on their own, not as a payload of another spacecraft like the Sojourner rover. After landing on Mars, these washing-machine-sized vehicles traveled over 2km on the surface. After happening upon an interesting rock, they would bring a suite of instruments to bear to look for the presence of past water on the surface. The instruments are on the end of an arm that lets them move into the best position to take any measurements. The first tool used is a microscopic imager, followed by the rock abrasion tool, or RAT, which is a fancy name for a drill. The RAT will expose a smooth surface of the rock, then look at the sample again with the field glass, then expose it to two instruments to determine what the rock is made of. The instruments (an Alpha-Proton X-ray Spectrometer or APXS and the Mössbauer Spectrometer) are designed to describe the makeup of the rock. The APXS looks for an overview of the materials, while the Mossbaer Spectrometer details iron-rich minerals. By mid February, both rovers landed successfully and were providing breathtaking images from opposite sides of Mars for a curious public, and amazing microscopic images and mineral reports for the scientists. All indications are that both rovers will operate for some time on the surface.

In 2005, there is a plan to send another spacecraft into orbit around Mars. The Mars Reconnaissance Orbiter, or MRO, will carry a larger camera than previous orbiters that will allow it to take images of items as small as a beach ball on the surface. The stated mission of this spacecraft is to scout out prime landing sites for future missions.

In 2007, a mission named Phoenix will travel to the north pole of Mars, looking for traces of life in the ice. The mission's name comes from the fact that it re-uses the Mars 2001 lander spacecraft, which was scheduled to fly, but postponed indefinitely after the twin Mars explorer crashes in 1999. There's no word out yet whether the build-up to launch will include a full-up system test, the kind that would have caught the error that doomed the 1999 lander, but good money is on the fact that there will be.

The 2009 plans are just starting to come into focus, but they involve launching a more-sophisticated rover than *Spirit* and *Opportunity*. Called the Mars Science Laboratory, the rover will be nuclear powered, giving it energy to operate 24 hours (and 40 minutes) in Earth time a day on the Mars surface.

The Challenge

Mars has all the raw materials necessary for a new branch of human civilization. We know this because we've sent spacecraft to the planet's surface, sampling the soil and air to find out. The challenge, simply put, is for us to get there and use them to expand our permanent presence beyond Earth.

Getting there will not be easy or cheap. No matter which financing method (governmental, commercial, hybrid, or other) gathers the resources here on Earth, or which mission design is chosen, the cost will be on the order of tens to hundreds of billions of dollars.

In my opinion, the simplest method to take humans to Mars was described by Zubrin and Baker in the early 1990s. Robert Zubrin used the Mars Direct plan as the basis for his book *The Case for Mars* in 1996, which sold well enough to inspire him to found The Mars Society, as mentioned earlier.

The Mars Direct plan is based on the assumption that humans can bring back a heavy-lift launch vehicle similar to the rocket that took us to the moon in the 1960s—the Saturn V. Using that vehicle, trips to Mars can take place using just two such launches each (or four, if the booster built has 1/2 the capability of the Saturn V).

A Mars Direct based mission starts with a spacecraft, called an Earth Return Vehicle (ERV), being launched to the red planet at a convenient launch opportunity. This vehicle is launched without a crew, but, as its name implies, it is designed to provide a crew's ride home. The ERV carries a propellant generation plant, a small nuclear reactor, and a supply of chemicals (likely hydrogen) for use on the surface. The chemicals brought to the surface of Mars for reaction are known as feedstock.

When the ERV lands, it deploys its nuclear reactor on a cart, which then rolls a few hundred feet away from the vehicle. The reactor starts and powers the vehicle to draw in Mars' atmosphere. Carbon dioxide, the primary gas in Mars air, is separated from the other components, and compressed. This compressed carbon dioxide is reacted with the feedstock brought along to produce fuel, oxidizer, oxygen, and water for the crew to use when they arrive. Depending on the production method (and the fuel) chosen, a Mars mission can generate 18 or even up to 54X the mass of fuel, oxygen, and water that they bring along as feedstock. There is even one chemical reaction, called the reverse-water gas shift[74], that can be put in a box and, as the unit pulls Martian air into the box, it can generate oxygen essentially forever by simply adding power.

While the production of oxygen and water is widely accepted as something useful to do, the produced fuel is up for some debate. Zubrin originally proposed methane, because it can be produced rather easily when combined with hydrogen in one chemical reaction called the Sabatier. Others have pointed out difficulties with methane[75], in that when it's stored as a liquid, carbon dioxide from Mars' atmosphere will freeze onto the tank, warming the fuel up appreciably. Other fuels that are possible that don't face the same storage issues are ethylene and benzene. These chemicals have the advantage of producing more fuel mass for each kilogram of feedstock, but it takes a more complex chemical reaction to produce them. To me, the fuel argument, based on whether or not the fuel is cryogenic (the term used to describe something that's extremely cold), is moot because the oxidizer, oxygen, is cryogenic. It's unlikely that another oxidizer will be proposed because oxygen also has the side benefit of providing breathable air for the crew.

Production of these chemicals, no matter which particular combination of fuel is chosen, will require a nuclear power source to work. While I've looked at other methods of powering these reactions, the high power requirements combined with the 24 2/3 hour day/night cycle on Mars' surface makes a nuclear power source the best option.

According to the Mars Direct plan, about 26 months after the first launch, the ERV is sitting on Mars' surface, fully fueled and stocked with supplies. Mission managers on the ground can verify this by looking at information beamed down from the craft. Assuming that all is OK, a crew launches to Mars on a 200-day trip, landing near the ERV. Detractors say that such a pinpoint landing is difficult or impossible, but precision landing was demonstrated in the *Apollo* 12 mission when Pete Conrad and Alan Bean landed 200m away from a space probe, Surveyor 3, that was sitting on the moon for 3 years before they arrived. Surveyor 3 was a dead craft, having shut down after completing its primary mission, but based on data we had about its location, Conrad and Bean landed in sight of it. In the Mars Direct plan, the ERV would be an active spacecraft, able to help guide its future human occupants to a safe landing right next to their ride home.

Once on the surface of Mars, the crew would plug their craft into the same nuclear power supply that ran the ERV, and they'd have all the water, air, and fuel that they'd need to research Mars in-depth. Their bulk food would come from Earth in a mixture of dehydrated and hydrated foods. The supplies of fuel and oxygen open up a wealth of exploration opportunities through the use of combustion-cycle rovers, which are much more flexible and have a longer range than their battery-powered counterparts. Rovers that run on turbine or internal combustion cycles can have ranges of hundreds of kilometers, and will allow exploration of huge swaths of land around the landing site.

One issue that any rover will face is heat dissipation. On Earth, the atmosphere is thick enough that a moderately sized radiator on a car will remove the waste heat from the car's engine. On Mars, the atmosphere is not thick enough to do that. Some other form of heat transfer will be required, either into the vehicle itself (a short-term solution); other options include dirt that's scooped into the vehicle or taking occasional rest stops where it transfers some heat into the ground. This is a design issue, but not one that's insurmountable.

Combustion-cycle rovers are not the end-all of mobility on the surface. It's true that nuclear-powered rovers will likely have even longer range than combustion-cycle rovers, and once the taboo of flying nuclear reactors on space missions is broken a portable reactor on a rover will likely be approved, but I think that, for now, the combustion cycle rover plan is better. There are also proposals for crews to use small rockets or "hoppers" for long journeys around Mars' surface. Many of these use fuels produced on the planet to minimize their need for supply from Earth. There's nothing fundamentally wrong with these ideas, but they're a little distant for discussion here.

One of the biggest research efforts in NASA for the last decades has been determining the impact that zero gravity has on humans. The goal of this research has been to find a counter to these effects, presumably in the form of some sort of pill or special exercise that would alleviate all the symptoms of zero gravity exposure, which are pretty severe. One way to avoid the affects of zero gravity has not been researched with any significance, even though it has the greatest potential of working. Here, I'm talking about avoiding zero gravity altogether. This can be done by generating artificial gravity on board the craft that people travel to Mars in. When I say artificial gravity, many people probably imagine the selectively available type that's used in most science fiction shows, the kind that can be turned on or off through some sort of magic under the decks, but that's not what I'm talking about. In the real world, knowing what we know now, we can generate artificial gravity by spinning our craft up. To do this, we need to bring a rope (well, ropes are called tethers in aerospace circles) and attach the rope to the booster that sent us on our way to Mars. Once we've done that, we use our small rockets to make us tumble end over end, and presto! All the countermeasures for zero gravity are unnecessary. This isn't as simple as it sounds, and some research and demonstration will definitely be necessary to show that it all works, but since this force is felt by anyone who visits an amusement park and rides a spinning ride, it's easy to describe as a concept!

Advantages

Mars' atmosphere is its biggest asset. While it is not nearly as thick as originally theorized before our probes traveled to it, Mars' atmosphere is thick enough to be tremendously useful, both as a brake for vehicles approaching the planet, and as a chemical resource for vehicles on the planet. Let's take a look at each.

Typical motion in space involves firing rocket engines of one sort or another. In order to burn these rocket engines, a spacecraft needs to bring propellants along. Each time an engine burns fuel is used. Unfortunately, in space this principle applies to both speeding up and slowing down. If you want to speed up you point your engine in the opposite direction to where you want to travel, and if you want to slow down you point your engine in the direction you're going to fire it. When you speed up, you need to carry the slow-down fuel with you as well, which makes speeding up harder. If you're going somewhere with an atmosphere, though, you can use the atmosphere to slow you down. There are two processes that use this approach. The first is called aerobraking, which has been done several times over two different planets. In this process, a spacecraft that's already in orbit around the planet skims the upper atmosphere. By doing this, the spacecraft slows down by an amount dependent on how deep within the atmosphere the craft goes. This method of using atmospheric drag to change orbits does not require any major changes to the spacecraft's design, but can take a long time (many orbits) to cause the necessary orbit change. The Magellan probe in orbit around Venus proved this concept, and then the *Mars Global Surveyor* and the *Mars Odyssey* spacecraft used it successfully to complete their missions.

While aerobraking is definitely a useful process, it has its limitations. The biggest drawback is the amount of time it takes for a spacecraft to change its orbit using the procedure. If you want to change your orbit quickly, you need to use the process known as aerocapture. Aerocapture in one form has been done before, because astronauts returning from the moon used the Earth's atmosphere to slow down for a landing. For Mars exploration, that type of aerocapture can be used, and has been used by the Pathfinder probe and by the Mars Exploration Rovers in January 2004. For human operations, however, where a more precise landing point (near the Earth Return Vehicle) is desired, it's likely that an aerocapture to orbit will be necessary. This method involves the spacecraft carrying an aeroshell, which is essentially a heat shield, and dipping deeply into Mars' atmosphere. At lower altitudes, the craft will lose a lot of velocity, and generate a lot of heat in the meantime.

After this close pass, the craft will swing back out of the atmosphere, into a somewhat egg-shaped orbit around Mars. Small thruster firings will prevent the craft from entering back into Mars' atmosphere, and the voyagers will be in a stable orbit. Later, the crew will fire small motors again, taking their path back into Mars' atmosphere aiming for the surface.

On their way to the surface, a crewed craft will get to use something never used in human spaceflight to reach a goal: parachutes. Again, Mars' atmosphere is not as thick as Earth's, so parachutes will not lead to the majestic scenes pictured at the end of the movie *Apollo 13*, where the crew hangs quietly under three beautiful canopies, but parachutes are more than worth their weight. Once the primary heating is over en route to the surface, parachutes will slow the craft down quite a bit, and allow landing to take place using very little fuel.

The other advantage Mars' atmosphere provides to space travelers is readily available resources. The Red Planet's atmosphere is made up mostly of carbon dioxide, with some trace gases of nitrogen, argon, and even water vapor. Carbon dioxide is more than 2/3 oxygen by mass, and the carbon atom is very useful, able to be converted through fairly basic chemistry into a number of different products. Water has obvious uses for a crewed mission, and the nitrogen and argon will serve a great purpose as buffer gasses, allowing a crew to breathe more than just oxygen on their journey. Such a buffer gas is not critical, but it does make some portions of the mission simpler. All of these resources are available to a crew who brings a supply of hydrogen with them, along with the equipment to do some 18th century chemistry. The Sabatier reaction was described earlier, as a way to produce methane and water from Mars' atmosphere, but the reverse-water gas shift, mentioned earlier as a way of producing oxygen, run on a larger scale and with different proportions of its reactants, has the potential of producing plastics for the crew's use[76].

Other solar system bodies, including the moon and asteroids, have some of the materials that I've mentioned here as being in Mars' atmosphere. Oxygen is found on the moon, and the asteroids as well, and we've proven pretty well that there's water on the moon (though its exact makeup—huge hunks or simply an ice-rich powder—is under debate) in some places, and some asteroids are definitely water-rich. In those cases, though, the minerals (oxygen and ice) are in the dirt underneath a spacecraft. When sitting on Mars' surface, a spacecraft is literally bathed in carbon dioxide. On the other bodies, mining (or at least shoveling) equipment is required, in some cases combined with a massive amount of power, to get at important chemicals, while on Mars, explorers would only have to open a window (figuratively speaking).

The *Viking* probes that landed on Mars found the planet's soil to be relatively rich in a lot of useful chemicals—phosphorous, sulfur, magnesium, all useful to industry on Earth are found in relative abundance in Mars' soil. Mars also has evidence of past (possibly even present, in geologic terms) volcanism. Volcanoes have a tremendous effect on the landscape and mineral concentration of a planet. With volcanoes, materials found in the crust can flow together into useful deposits, making for good mining possibilities. Plate tectonics, present on Earth giving us our changing surface and giving California its infamous earthquakes, are also useful in this concentration, but it's looking more and more like Mars never had plate tectonics, or if it did, they stopped so long ago that there are few traces today.

It all boils down to some simple facts. Mars has all the materials necessary for a human contingent to make a lot of the materials they need to live. Some of them are easily accessible by processing the planet's atmosphere, while others will require more work in the form of soil processing or prospecting and mining.

Disadvantages

Philosophically speaking, there are those who say that a mission to Mars would do nothing to enable spaceflight overall. They compare a mandated Mars mission to the mandated Apollo mission that sent a few very expensive spacecraft to the surface, produced no tangible benefit, and then got canceled. The argument continues that the Mars mission would not require a huge increase in launches off the Earth surface, and therefore wouldn't cause a shift from government-funded to commercially funded space. Not only that, they say, look how far you have to travel!

Distance

Mars is a long way off. In August 2003, it came the closest to Earth that it had in tens of thousands of years, and it was still 52 million kilometers (34 million miles) away. Our trips to the moon took three days each way, and it's only 380,000 kilometers (about 1/4 of a million miles) away. Unfortunately, given today's technology, and the technology reasonably attainable in a 10-15 year period, we can't go directly to Mars on a close approach, either. Instead, our Mars travelers will have to take a looping trip that will take them almost half way around the sun on their trip, which translates to a much longer trip than the minimum distance, and means that the trip will take a while. How long is a while? About 200 days, depending on the amount of energy you can expend. If you can expend a little more, you can shorten the journey a bit. I should note that this is not the minimum energy trip, which would translate into about 25% more time on the way, but I'm bringing that up mainly for the folks who are reading this book to watch me drop something technically so that they can say I don't know what I'm talking about. The longer-distance option is a good idea for cargo missions, where time spent traveling is not as big a deal. By taking the longer route, a cargo craft can deliver more mass to the surface as well.

While people on Earth say time is money, in space communications, time is distance. Because Mars is so far away, normal communications as we know of here on Earth will not be possible. Radio signals, traveling at the speed of light, take anywhere from five to twenty minutes to travel from Earth to Mars. The longest delay humans have dealt with in communications so far took place when we had people in the vicinity of the moon, but that delay was only on the order of a second. Because of this, a Mars crew will be on their own more so than any other recent human endeavor. If some sort of disaster befalls them, the ground team won't even know it until the necessary amount of communications time passes. By that time the crew is either well on the way to fixing the problem or the ground team will be unable to do anything about it.

Some Mars mission architectures put a crew of 4-6 in a "tuna can" habitat for their journey. Small, but functional, individual quarters are a must, and the total living area on board the craft would be on the order of 78 sq m (825 sq ft). For some people, that fact alone is a show stopper. Many of the people of this opinion are either those who would never make the trip anyway or are psychologists with a lot of grant money riding on research on the psychological effects of isolation. Articles written on the topic relate scenarios like "Imagine traveling with your family of four in a small Winebago that you can't even walk outside for 7 months...." While the analogy brings a smile to my face, I don't put a lot of faith in it. Any research into the great voyages of the past shows that great people with a goal in mind can put up with a lot of discomfort on the way to their goal. While conditions aboard the tuna can to Mars may be considered unlivable by the likely mid- to upper-income readers of *Wired*, I venture to say that there are others (myself included) who would volunteer to go on such a mission, and that the opinion of *Wired* readers about travelers' living conditions shouldn't stop volunteers from going.

Mission Length

I've heard a mission to Mars described as a non-starter because of a simple fear of mechanical failure. A crew that leaves Earth for the fourth planet will be away for almost three years, and some say that entirely too much can go wrong in that timeframe for us to risk the space program on it. Again, based on what we've done so far, building space stations that require constant support and supplies from the ground, I can see where that conclusion comes easily, but we haven't tried to do anything else yet. Once some sort of Mars program comes into being, there will be many demonstration flights of increasing duration, improving our skills at building craft that can last three years. These first flights won't travel the whole way to Mars, but will instead stay in the near-Earth environment (likely in high Earth orbit, such as the L1 point between the Earth and moon, described earlier). This type of shakedown cruise has some drawbacks, in that it can't model the day/night cycle of Mars, but that can be done on Earth. Mars dust (more on that in a moment), always a concern for Mars mission naysayers, can be simulated (made on Earth to match the chemical makeup that we've measured Mars dust to have) on board our checkout craft, being introduced into the system in doses that far exceed expected amounts. I imagine a pretty comical scene on board a test craft where the commander walks out with a big plastic bag of Mars dust and dumps it into the ventilation system. Then, if there's a problem, the crew can quickly return to Earth to have the issue repaired and re-demonstrated on another mission. Over time, people should grow to see that we can rely on machines for years without repair and resupply from Earth.

Contamination (Forward and Backward)

Mars is one of the places in our solar system thought to harbor life (we know Earth does, other candidates include Jupiter's moon, Europa), and this provides a couple additional complicating factors, known as forward and back contamination. The question of forward contamination goes something like this—if there is life on Mars, and microbes that our crews take with them on their mission destroy that life, what should we do? Back contamination, as its name implies, deals with people who've explored Mars (or robots who've explored Mars) returning from there with diseases, or just other life forms, that life here on Earth can't deal with. Opinions vary wildly on these topics from the crassly dismissive to the overtly alarmist.

In forward contamination, the varying opinions swing from "if there's life on Mars, and it came about separately from Earth's, we need to leave it alone, or at least minimize our impact on it" to "if we're going to say we won't go to Mars because of our impact on the bacteria there, we may as well outlaw mouthwash." One particular advocate of the former statement, Chris McKay of NASA Ames Research Center, points out that it's possible that life found on Mars may be entirely different than life on Earth. In this "Separate Genesis" scenario, Mars life is different enough to rule out a common source of life for both planets. One particularly strong indicator of this would be DNA strands that rotate in an opposite direction from ours. In such a case, proponents argue that scientific study of the "other way" to have life outweighs any advantages that would be gained through colonization, mining or otherwise exploiting the Martian landscape. I'm not sure that I buy this, because any life that exists on Mars right now must be incredibly hardy, or buried incredibly deeply in order to hold on. In either case, I think that humans would be hard pressed to alter the Martian environment enough through terraforming, or have our alterations reach deep enough for our presence to impact the life forms there. Either way, we're probably some ways off from reaching a decision point on this, because we have to get to Mars first and find the second genesis before the debate turns anything other than academic. The other extreme in this opinion, voiced by Robert Zubrin, basically says that we don't let bacteria on Earth (the likely form of life found on Mars if there is any) stand in our way right now to the point that we have an active, if unsuccessful, effort to destroy it all. If you don't buy that argument, check the bragging rights of anti-bacterial soaps. Would you buy one that claims to wipe out 95% of germs over one that claims to wipe out 99%? This "mouthwash" opinion strikes me as extreme also, but it remains academic until we actually find something.

The ideas of back contamination deal with Mars life coming here to Earth and causing problems. These problems range from no impact whatsoever to the end of life as we know it on our planet.

The former argument, that potential life forms on Mars returned to Earth would have no impact whatsoever, is based on some facts, but championed by many without a biology background. The argument goes that Earth and Mars exchange material all the time in the form of meteorites. One direction of this exchange has been proven, in the case of meteorites like ALH 84-001, for which there is no scientific debate left that this rock came from the planet Mars. (There is significant debate remaining as to whether or not the rock shows traces of ancient life on Mars.) The other direction (Earth-Mars) exchange of material is theorized and highly likely, but unproven at this time. Because of this constant exchange of material, the planets have already dealt their worst to each other. That is, any bacteria that exists or ever existed on

Mars has already made the journey to Earth and caused its harm. Another portion of the argument goes that even if people on Mars discovered something new, the new bacteria could not infect Earth life because of the way diseases and their hosts develop together over a long timescale. My personal favorite, although overly simplified, argument along these lines was posed by Dr. Zubrin, where he states that sharks are not a threat to lions as the top carnivore in Africa for obvious reasons. Essentially, diseases and their hosts have co-existed for years, and have learned to attack and defend from each other. The third portion of this argument goes that humans on Mars, upon discovering some life form, will be isolated from the rest of human civilization for more than 500 days. If a problem doesn't develop in that timeframe, odds are the biosphere on Earth is safe.

While I tend to agree with these arguments, I don't have enough education in biology to make a knowledgeable decision about their validity. In the course of e-mail discussions with researchers who hold the opposite opinion, I've gathered that the following arguments summarize their concerns.

First, on the assertion that planetary exchange over the years has inoculated us from any potential "bad bugs" from Mars, the other side of the coin states that we don't have a good feel for where new diseases come from now. Plague was cited as one example of a disease that wiped out huge numbers of people, and no one knows where it came from. Therefore, the argument continues, it might have come from space, either Mars or some other rock, and we can't run the risk of bringing a plague back to Earth. In a counter to the 500-day exposure period on Mars proving that there are no pathogens that can harm humans, the response asks, "What about all the other animals? What if humans on Mars pick up a disease that doesn't affect humans, but wipes out all the honeybees?" The primary focus of this opinion is focused in a group called the International Committee Against Mars Sample Return (ICAMSR). ICAMSR is discussed in an earlier chapter. When last I checked, though, the ICAMSR website was still operational, and I'm sure the founders would appreciate an e-mail to show that someone is still reading their stuff. The recent crash and contamination of the *Genesis* return capsule, an attempt to bring material back from space, should provide plenty of fodder for them.

The ICAMSR arguments strike me as a little extreme, but again my background is not in biology. So, I'll take the cheaters way out and say that the answer is probably somewhere in between. Unfortunately, any type of isolation for a returning Mars craft will be difficult, especially a human craft, given the dymanics of how that ship will very likely approach Earth. Because it's much easier to have a small craft plunge into the atmosphere (either to return to the surface or just to slow down into orbit), the craft will have very few supplies to support the crew immediately after that trek into the atmosphere. The Apollo missions used this approach, returning their crew to Earth in the tiny capsule, which was excellent at getting through the atmosphere, but not very good at supporting its people for a very long time. If a returning Mars craft is expected to use the atmosphere to slow down into orbit and then stay there for some time, it will pose a significant design challenge. If the craft remains large enough (by carrying fuel on board) to fire rockets to enter Earth's orbit, then the mass of the entire craft may be hard to build to. In the end, a return mission will likely come back to an isolated area (think desert Southwest of the US, Siberia, or the Australian Outback) or they'll land in the ocean and won't get out of their capsule until they're isolated on board ship.

Radiation

Another detractor often brought up as part of a Mars mission is radiation. We've discussed it before, but Mars exploration brings some unique aspects to the table. Radiation is a byproduct of atomic energy, one of the most powerful forces humans control today, but it's also part of our everyday environment. As part of its connection with the former, it's had a difficult public relations run (rightfully so in some cases to be sure, such as the survivors of Hiroshima, Nagasaki, and Chernobyl will testify to) but it's also served humankind in many ways, such as x-rays, allowing us to understand ourselves better. Continuing on its bad PR trend, radiation has been brought up as a "show stopper" for a human Mars exploration. We don't hear a lot about radiation in space travel today because current human activity is in low Earth orbit exclusively. When in low Earth orbit, the Earth's magnetic field protects a craft and its crew from a significant portion of space radiation, but if we're going to travel beyond LEO, we'll lose this protection. Let's take a look. Part of this will be review from earlier, but I think it's important to review pertinent details as part of the discussion.

You will remember that radiation in space, the kind our travelers will face in transit from Earth to Mars and back, is made up of two types. The closest-to-home form of radiation comes from our sun, and we know which direction it's coming from at all times. This type of radiation can be shielded against using on-board ship stores (water, food, and human waste) to create a "shadow" of protection for the crew during the worst solar events. Luckily, these solar events are short-term, lasting on the order of hours ours days. The

217

11-year solar cycle creates a mirror cycle in the other form of radiation discussed, galactic cosmic radiation, or GCR. GCR is a much tougher foe to fight, and at this point there is no demonstrated method of combating it effectively. These two forms of radiation characterize our crew's radiation hazard for about 400 days of an approximately 950 days Mars journey.

Radiation on Mars' surface is another matter, and I wanted to address it separately because there's been a lot of incomplete or just bad information going around on it. Most of the information I'm referring to is the data released from the 2001 *Mars Odyssey* orbiter that entered Mars' orbit in October of 2001. 2001 *Mars Odyssey* carried an instrument on board called MARIE, and it was designed to measure the radiation environment en route to Mars, and around the planet. In May of 2002, in a press release that was picked up by most of the major news sources, NASA went on record saying that the radiation environment on Mars was hazardous to humans. They went on to say that the radiation environment on mountain peaks of Mars was worse than in low-lying areas. Very little perspective was offered in these pieces, except to say that astronauts are also exposed to radiation on the ISS.

First things first: it is true that Mars doesn't have a magnetic field and that its atmosphere is much thinner than Earth's. This means that the radiation exposure a person would receive standing on Mars would be greater than the radiation exposure a person would receive standing on Earth. These facts are no-brainers. It's when you start comparing this amount of radiation to the amount of radiation that a person would receive if they are not on the surface of the planet that things get a little distorted. First of all, the planet itself serves as an excellent shield for radiation. If you are standing on Mars, fully 1/2 of the sky that would be visible to you in space is blocked by the planet, so you're better off already. Remember that GCR comes at us from all directions evenly, so the planet blocks 1/2 the GCR radiation at the start. Mars' atmosphere blocks some of the sun's damaging rays, but not enough to make us really comfortable, and as mentioned before, the cosmic rays are really hard to stop. So, what do we do? The answer is as easy as dirt. In fact, the answer is dirt.

While we can't build spacecraft with enough shielding to protect a crew from the dangers of cosmic rays on the journey to Mars, it is within our capabilities (though we've never done it) to build a spacecraft that can have sandbags stacked on top of it. Dirt is a pretty good insulator, and the more of it you have, the better off you are. Luckily, we don't need to bring this dirt with us, because it's all around on the surface of Mars. With enough dirt on top, a Mars lander can protect our crew from some powerful cosmic rays and all but the absolute worst solar storms. As another safety step, outdoor activities can be limited to early morning and late afternoon timeframes most of the time. Then, our explorers are going to be in a pretty benign radiation environment.

The 2001 Mars Odyssey results can't be ignored in all of this. Yes, it would be better for a mission to land in either Hellas Basin (the lowest land on Mars) or in the northern lowlands. Landing sites on the top of Mount Olympus (named for the home of the gods in Greek mythology, it's the tallest mountain in the solar system, and its peak is above a large portion of Mars' atmosphere) are not recommended. In fact, trips to the top of Mount Olympus would be somewhat anti-climactic, because the slope to the top of the huge volcano is walkable without special equipment all the way up.

In summary, Mars radiation is an issue, one of the most serious facing such a mission, but the data we have right now, combined with some common-sense actions carried out on the surface, make the risk manageable. Research has shown that a man making the journey to Mars and back would increase his cancer risk for the next 30 years by 5%, and a woman would increase her cancer risk by slightly more.[77] These numbers are not to be taken lightly, but they're not a reason to pass up on a great human adventure that has the potential to preserve our species, either.

Dust

Dust is another difficulty a Mars mission must face, but so must any mission that lands on another planet or moon or asteroid. Martian dust is a mixture of silicon dioxide and iron, with some other materials thrown in. In many ways it has some similarity to Earth sand, but there are a couple outstanding features that make it much more difficult to deal with. First is the fact that Mars is so dry. Without long-term exposed water on the surface or condensing water in the atmosphere, the dust is left either suspended in the air or deposited on the surface in a very fine layer. The slightest puff of wind will lift this dust off the surface and carry it for tens if not hundreds of kilometers, coming to rest on the surface again or on any human-made object that happens to be there. This brings up the second distinguishing factor of Martian dust, its reaction to magnets. Iron is a major component of Mars dust and its sand. In fact, the iron is responsible for giving the planet its color and thereby its name. Having a magnetic component in the dust creates a problem, however, in that anything exhibiting a magnetic field (electronics, motors, most of the things we take for granted here on Earth) will attract this dust like crazy.

Lastly, the dust found on Mars is very fine. It started as larger grains, but over billions of years of meteor impacts, wind exposure and potential fluid events, it's been broken into smaller and smaller pieces. It's currently estimated that some of the dust is so fine that it will pass through most filters that we have today, and if allowed to get into a crewed spacecraft (I see some dust in the craft as a certainty anyway, but there are those who maintain that it can be kept out) it will pass through the crew's sinuses and directly into their lungs. There, the dust could be anything from an irritant to a cause of major disease to a short-acting poison. At this point, we don't know because previous exploration of Mars' soil was incomplete. We know some of its components, but what remains can only be theorized, and the worst scenarios are normally the ones that draw attention.

When discussing Mars dust, a parallel is often drawn between it and the lunar dust our astronauts were exposed to during the Apollo era. While comparisons like this can be useful, we have to be careful to use the pertinent portions of the comparison, while ignoring the parts that don't apply. The dust on the moon did get everywhere, as anyone who's visited the Air and Space Museum can see when he or she sees Gene Cernan's moon suit on display. The gray powder of our nearest neighbor, however, has only been shaped through cosmic impact. There are no winds on the moon, and no fluid motion to round off the extremely sharp edges produced when rock is shattered into the tiniest shards. This lack of polishing makes the dust extremely damaging, working its way into any mechanical joint and tearing it apart.

Lunar dust was not free of extreme theories, either. Some scientists believed that the moon dust would burn or explode as soon as it was exposed to oxygen. This wasn't a problem for astronauts going out and walking around in, but I'm sure that some of these scientists were holding their breath when the explorers returned to their lunar module and re-filled the capsule with breathing air. Of course, once the cabin repressurized and the theorized explosion didn't happen, the people who strenuously pointed out their theory were hard to find for quotes.

So dust is something we'll have to contend with, but it's something we've dealt with before. The differences are the magnetism, size, unknown chemical components, relatively eroded edges and length of exposure. Bring on yet another challenge.

Human Factors (Again)

Human factors are another issue that people bring up against missions to Mars. To summarize, the argument goes something like this: "People can't be isolated in small groups for a long period of time without running into problems. Also, it's hard to say what kind of effect being so far away from Earth will have on people."

We've discussed human factors before in these pages, and I believe that they are overblown. Those who are not interested in making any journeys themselves propagate most of the assertions made by the psychiatric community.

NO DESTINATION, JUST REGULAR ACCESS

During the space policy debate of 2003, there's been a small, vocal group advocating no destination at all right now. Their fears relate to a monolithic space effort, similar to the Apollo program of the 1960s, crowding out all other space activities. While the specifics of their ideas vary, their basic goal remains the same: make more of humanity spacefaring.

The Goal

Space hotels and relatively cheap flights that anyone can buy like airline tickets are what commonly comes to mind when people think of easy access to space, but visionaries look beyond that. With a drastically decreased cost to orbit, space suddenly becomes a reasonable destination for any entrepreneur with an idea and the manner of gathering enough cash for the project. The impact of a money-making effort in space would have an enabling effect on the industry, and cause an explosion in access methods and available hardware for use in space.

Large groups of people who pool their money could hire one of the space hotel builders to build them a small colony, and humans could move beyond Earth to live. As Robert Heinlein stated, once you're in low Earth orbit, you're halfway to anywhere, so colonies in orbit would be followed quickly by colonies beyond Earth orbit. These colonies will need supplies, requiring more flights into space, and the cycle repeats itself.

The Challenge

All discussion aside, the fact is that this is just not how space is done today. As it stands right now, space is primarily used by world governments, with occasional licensing granted to companies with something to offer one or more of those governments, either in the form of imagery, communications, or simply prestige. Access to space is controlled even tighter, with most launches taking place out of government-maintained launch ranges, on vehicles that were wholly or partially financed by government development contracts. Any government's job is to control things, and they typically don't surrender control of things lightly. In order to crack the current cycle, there will need to be some sort of demonstration of another system that works. There are those in the space advocacy arena who believe that many such opportunities arose, but that they were crushed by a government's action in desire to maintain control over space activities. I don't buy that completely, though I think many people underestimate bureaucratic inertia. I think that many revolutionary space activities fell ill due to a combination of government disinterest and internal problems. What's needed to crack this cycle is for a new space venture to work, to provide a proof-of-concept.

The proof-of-concept doesn't need to be big, it just needs to work. Whether a start-up company produces an economically viable rocket that can fly on demand and deliver its payloads as requested (SpaceX?) or a space prize spurs someone into doing something radical (the X-Prize) it just needs to work. After one breakthrough exercise, it's likely that there will be a partial swing in public interest in space. Once the space arena is seen as something which, though it remains difficult, is not purely the domain of governments and their contractors, people can see themselves involved. People can see themselves in space.

Advantages

With multiple groups working on multiple methods of reaching orbit, while other interests develop life-support technologies, power systems, and living facilities for the space environment, the entire effort is not reliant on the actions of one particular party. This is how things happen in most sectors of business today, with suppliers jockeying for the best position based on cost and quality while a group of integrators are always developing new projects to sell, keeping the whole system going.

A hypothetical example of this sort of action in space is a little tough to come up with, but imagine if the International Space Station was built with enough flexibility that the next part that was built could be the next part launched. Nations would scramble to have their part ready in time to be the next piece. As it exists today, the station requires assembly in a very specific order, and one critical piece, the Russian Service Module, held up construction for months while it completed final development and testing. The concerns were so great on the part of the US space agency that they actually researched building a replacement module for the station, though in the game of international projects and politics, it's hard to say how serious the effort was. It may have just been a ploy to spur the Russian Space Agency into action.

Disadvantages

I've quoted Jeff Foust before in saying that the best rocket fuel is money, and money is definitely the biggest problem with this theory. While some companies are building small portions of this infrastructure with their own cash, a space effort, at least in the past, involved global efforts of communications, as well as recovery and emergency planning. Granted, it's possible that new methods (landing at the launch site, satellite communications, global positioning) will cut down on the needs of the past, but the arena of spacecraft construction (including hotels and passenger spacecraft) will remain a significant hurdle without a major paradigm shift.

The size of these efforts and hurdles translates directly to money, and money doesn't come unless there's some sort of demonstrated return. This is where I feel there's a great opportunity for a hybrid of government and commercial activities, with a government that's more interested in the results of an action (payload delivered to orbit) than the technology that made the action possible.

While I won't rule out some business interest, I will say, as I've said before, that it's unlikely until there is a demonstrated success. Here, an incremental approach will serve well, with small projects building to larger ones as time goes on. This is the approach favored by Elon Musk in his SpaceX venture.

Right now, at least, space has a "barrier" problem in that, based on previous experience, it is extremely difficult to get something into orbit. The barrier was even quoted in the Columbia Accident Investigation Board report, citing how difficult it was to place a space shuttle in orbit and return it safely, even though may people consider the action routine because of how many times it has happened. Like other barriers, it looms large right now because we haven't found a low-cost way to break through it. I believe that sometime in the future, the Earth-surface-to-orbit barrier will hold a similar standing with the sound barrier, something that is still there, but can be overcome with proper care and skill. For now, though, that barrier looks quite formidable.

Beyond

Things get a little hazy from here on, and I don't really dwell on missions beyond Mars in my day-to-day thoughts. It's true that main-belt asteroids (between Mars and Jupiter) have riches almost beyond comprehension, and that the moons of Saturn and Jupiter (likely in that order) are excellent sources of helium-3 for our supposed fusion reactors, but things change once you start traveling that far. Given the facts as we understand them now, including such things as the intense radiation fields around Jupiter, and the extremely long travel times to the outer planets, I think we'll save these discussions for another day. If you're interested in a fairly thorough look at them, I recommend the book *Entering Space* by Robert Zubrin.

CHAPTER EIGHT

FUTURE TENSE—HOW IT COULD HAPPEN

Having covered the possible destinations, our discussion turns to the methods of reaching those destinations. There are as many methods and variations on their themes as there are places to go. Of course, groups of people have strongly held opinions on which method must be followed. Let us try and take a dispassionate look.

SCENARIO 1: ANOTHER APOLLO

"First, this nation should commit itself to achieving the goal...before this decade is out of landing a man on the moon and returning him safely to the Earth."

There are people within the space community who believe that this sort of challenge for the next step in human exploration of space is the way things will go. Kennedy's challenge of 1961 focused the entire United States military-industrial complex on luna as a goal. The folks who see this type of model as working again in the future believe that it's the only way that exploration will move outward.

Preamble

Kennedy's challenge to the nation was not made in a vacuum, nor was it made for the glory of the achievement. The United States had suffered a series of humiliating defeats in recent years. In 1957, the Soviet Union launched the first artificial satellite, and then in 1961 they achieved another first by launching the first human into orbit. On economic and military fronts as well, the Soviet Union appeared as a threat to the United States, and a race to the moon was a peaceful way to show our technological strength. So, what kind of challenge today would spur this type of response?

The first that most people think of is a space race similar to the type going on during the '60s. In the current world environment, the United States is viewed, for better or worse as the sole superpower (personally, I like the term hyperpower) on the globe. So, any nation that wants to challenge the US in such a race would have to appear as a threat across a broad spectrum. It's true that China is moving towards a manned presence in space, becoming only the third nation to achieve that stature, yet not too many people, aside from my dad, view the nation as a serious threat on the economic or military fronts. So, for China (or any other nation, for that matter) to be the cause of a new space race, either their stature on the world stage must rise to the point where America perceives them as a broad threat, or the United States' stature must fall to the point where such a competition is seen as something worthwhile. If I've not betrayed it in my prose, I don't see this as likely in the near future.

The mechanism I see best at work here is the desire for safety. The survival reflex is one of the strongest of any creature, and humanity, despite its bureaucracies and silliness, hasn't buried its survival instinct yet. I believe that the discovery of a near Earth object or an inbound comet that will strike Earth will invoke this survival reflex. This was discussed in the asteroid section of "Expansion Outward," and I'll say it again—I hope that we don't come up as big losers in the waiting game. I'd much rather do an in-depth survey and know that something is coming in ten or twenty years (preferably more) than have to throw something together in just a year or two after finding something heading straight (orbitally speaking) for us. There's another possibility that a small asteroid could impact Earth, causing some sort of regional or coastline calamity, and that humankind would then take the threat seriously, starting an in-depth survey and planning for future impacts.

This all boils down to the fact that humans haven't proven themselves to be great long-term planners. With few exceptions, governmental plans are reacting to current events instead of planning for future ones. In some cases, one event will cause a reaction to prepare for future ones. The most recent example of this reactionary position is the war on terrorism waged by the United States and its allies. Whether you agree with the war or not, I believe there's very little doubt that it wouldn't be happening if September 11, 2001, had been a normal autumn day.

Despite the impending-disaster feel behind the whole scenario, it's rather easy for me to imagine a future president (or, if you're feeling really global, secretary general of the United Nations) standing before a hushed crowd of delegates after the formal announcement of an impending impact stating: "Humankind will not go down without fighting, humankind will survive, as we have in the past. We will face this challenge and beat it!" The crowd goes wild, and few people question why more wasn't done before....

Mechanism

Assuming that some sort of Kennedy-style challenge is laid down, a line of people who've been less-than-fully utilized will be ready to take action. They'll dust off old procedures and likely not work to update them too much. Humans will take on the challenge that they face, but how?

First of all, the floodgates of federal spending, likely from nations around the world, will kick in. Depending on timelines, that is, whether there's a space rock headed towards us right away or not, the levels of spending could make the Apollo program look like a minor effort. There would be little dissention in the congressional budget process, as is usual when something really important comes up. NASA centers, long underutilized or perhaps closed, will ramp up activity to unprecedented levels. New buildings will be built to take on the new organizations required to carry out this new level of homeland security. Development programs, long thought to be wildly expensive or just too crazy to work, will get funded, but the impact doesn't stop there.

Military contractors, who'd long let their space divisions shrink or vanish all together, will scurry to meet the demand in new spending. It's possible that some large-scale facilities for building commercial aircraft or fighter planes will be converted to a spacecraft factories, although the argument that will likely carry the day is that it's cheaper to build something new. Once again, the timeline for action plays a large part in these decisions.

In a matter of days, it will become "cool" to work on the space program again. During the Apollo era, it wasn't unheard of for workers in spacecraft factories speeding to work to be stopped by police, and once the officer knew who the person was and where they were headed, he or she (who are we kidding. In the 60s an aerospace worker or police officer was most likely a man. Today, however, there are many more women working in both fields) would provide a police escort and speed even faster to the destination. Those types of stories would be in vogue again. Of course, the work being done would be important on a scale unimagined in humankind, and it would be unclassified unlike the Manhattan Project to produce the first atomic weapon.

Space societies, as they're known today, will likely suffer a decrease in followers. To many people, the goal they were trying to achieve by joining the societies will be met. Given my experience with human nature, those societies will turn up any rhetoric they've had going talking about how the government is doing things the wrong way, but the public at large will not listen.

If our warning time until impact was greater than a couple years, enough time to get a generation of children through school, the face of United States education would change radically. All the NASA programs devoted to make more people interested in math and science would be unnecessary (given the staying power of most bureaucracies, they wouldn't go away, mind you, they just wouldn't be necessary) because kids would know how important working in those fields would be. This type of reaction has taken place twice in American history. After the Soviets launched Sputnik, entire school curricula were revamped to include more math and science, and a group of kids including Homer Hickam grew up wanting to work in the space field. In the other case, I speak from personal experience when I say that the Apollo program inspired me to study aerospace engineering and become the person I am today.

In rather short order, entirely new hardware will start rolling off the assembly lines. There will likely be a new class of heavy-lift launcher like the Saturn V, designed to carry huge pieces of spacecraft into orbit for assembly and redeployment in defense of the planet. The normal hand-wringing that goes with sending humans on board a spacecraft will be swept away, and those who travel in space will be true heroes, because their daily job will involve saving the human race.

News programs will change their focus, for at least part of the time. It's likely that whenever the impactor is being discussed, a clock on the screen will count down the time until the strike. After the initial rush of news, when the space systems fall into general development, there may not be a news story every night, but when it does come up most people within earshot will perk up a bit. I can't consider it likely, but I do hope, that in such a time of crisis the normal entertainment reporting would fall in stature a bit. Who cares which movie starlet married which director when the planet is on the line!

Advantages

This scenario is not without advantages, the primary one being speed. As the United States proved in the 1960s, once a goal is set and resources are available, even seemingly insurmountable goals have a tendency to fall by the wayside. The centralized planning efforts of an agency such as NASA are not perfect, but they will pull an action together faster than a loose conglomeration of activities from entrepreneurs. I doubt, however, whether a government agency could beat out a tight-nit group of entrepreneurs or a single, money-laden one.

Disadvantages

Rebuilding a space effort through an existing government bureaucracy is not an efficient prospect. This effect is compounded when that agency, for right or for wrong, has faced a period of fiscal neglect. The Columbia Accident Investigation Board went out of its way to point out the crumbling infrastructure that NASA works in, and at least a portion of a sudden surge of money would be used to update that infrastructure. The government purchasing process does not have the best reputation for cost controls, either. Those who oppose space activities are quick to point out how expensive it is to build space hardware, and how costs often rise, but this is a common occurrence in many government purchases. How many military procurements can honestly say that they are on time and under budget? The fact remains that doing new things (like traveling beyond low Earth orbit) is expensive, and when done by an organization with potentially unlimited funds it will be more so. Funding isn't the only issue that detracts from this plan, however.

Publicly funded activities are subject to a change in public opinion, evidenced by the rapid retreat from the moon after our initial successes. There are a number of factors cited as the reason that the public lost interest in the moon landings. The war in Vietnam is a favorite, as well as a goal achieved, the Soviets beat, and the change in presidential administrations. These were all players, but there may be something a little deeper. In the late 1960s and early '70s, coincident with the counterculture movement grew a creeping distrust of all things huge and governmental. There aren't too many examples of things that are huge and governmental that people are willing to live without, but the space program may be one of them. Politicians found that the space program couldn't be completely disbanded because of the employment it provided in key voting districts, so the compromise involved an agency given a project (the space shuttle) with enough funds to develop it and keep it in some form of function.

There are groups and individuals who say that the government has a stranglehold on space access, and that the government will do anything to maintain it. Those who make this argument point to a number of attempts by private industry to provide launch services and note the fact that none of them have been successful. While some of the plans were completely out of line, either in their business case or technical approach, others had a feasible technical approach but failed due to a single bureaucratic decision not to use their services. Of course, reliance on one decision like that could be construed as a bad business plan, even though there are notoriously few customers looking for reliable access to space within certain constraints such as mass and orbit location.

Well, a massive, government-funded effort to send people beyond low Earth orbit would not diminish those concerns. I've heard arguments that a new government focus beyond LEO will free LEO up for private action, and that may be the case. One thing to remember, though, is that the US military, actually a larger use of launch services than NASA, will retain its focus on Earth orbit even in the event of a large program moving humans beyond.

Personally, I believe that the government's control of space access has grown more out of necessity, lack of vision, and some bad business decisions than any active effort on the government's part to maintain it. I consider the Air Force's attempt to rely on the commercial market to lower costs in the EELV program as an effort in the right direction, and that it probably would have worked in many ways without the collapse of commercial space programs that we've discussed.

I read in an editorial that the difference between government activities and commercial activities can be summed up like this: In government, the goal is to do nothing illegal. If nothing happens, then by definition, nothing illegal has happened. The commercial world, however, wants to get things done and worries about the law later. I find this argument interesting, if simplified. A government agency is rarely called to task for inactivity. If budgets aren't being broken and no lives are lost, an agency could carry on for years without any undue concern. Once an agency starts doing things, like researching new materials (requiring money), sending people into space (exposing them to risk) and the like, that opens the agency up to criticism in the form of newspaper editorials and congressional hearings.

Unfortunately, government programs are largely a zero-sum gain—to win funding for one project, another project has to suffer. The space program is often cited as a money sink for funds that could do much good elsewhere. Under the proper conditions, however, a government shows amazing flexibility in funding new initiatives without removing funds from other programs. Examples of this kind of flexibility include the savings & loan bailout of the late 1980s (estimated by some[78] to cost $175B) and the ramp up in defense and homeland security spending after the September 11th attacks (still counting…). Both of these examples had serious economic or security concerns for the nation, which a space effort lacks, but they demonstrate how easily money can be found when the desire for it is present.

My Opinion

While I maintain that I will support any effort to take us deeper into space for the long term, I don't believe that a purely government-funded program is the answer. I believe that government has a role to play in such expansion, either in a hybrid arrangement that I'll talk about in a bit, or a government agency like COMSAT, that's designed to turn into a commercial venture over time. Pouring more money into an existing bureaucracy is not the best answer for taking humans into space to stay.

Scenario 2: Free Enterprise

"Let free enterprise take us to space!" This call is as pervasive in some space communities as the Kennedy nostalgia is in others. To hear two members of the opposing camps talk, you'd think that the goals were entirely separate. Never mind the fact that both methods would get people into space, and would provide basic technology to allow others to follow.

Preamble

It's unlikely that a purely free enterprise system would start through the same type of pivotal event that would jump-start a government-only space venture. Because of the inherently chaotic progress of free enterprise, it's more likely that a group of entrepreneurs would be working on separate projects, and eventually reach a point where they realize that, if combined with a new focus, their efforts could take people into space. One of the difficulties of this system is that each effort would have to be self-supporting before they all came together, because if they weren't they'd rely on some sort of governmental support, and would lose their distinction as a free enterprise project.

Mechanism

There are some common approaches cited as ways that free enterprise can get us into space. Each of them has some strong points, but there are also some problems with the simplistic plans.

Advertising Revenues

First, there's the theory that advertising revenues can finance a mission to space. This is a personal favorite, because I've had some heated discussions over the topic with one of its proponents. The argument goes something like this: "If the Olympics can raise millions of dollars to fund buildings in an Olympic village, and network television can sell rights to advertise between events for even more money, then it's possible that more money can be raised for a mission to space, which will gain even more of a viewing audience and be on the air for years." For example, a Mars mission, as currently planned, will last between 2 and 3 years. In some ways, it sounds plausible that such a long event could draw a lot of advertising, but the reason that the Olympics can garner so much money is that they're a proven revenue source. This wasn't the case before 1984, when Peter Ueberroth changed the thinking of how Olympics could be run during the games in Los Angeles. Building an Olympic village of living quarters, while difficult, is not necessarily cutting-edge. Plus, the turnaround time from investment to payoff is relatively short, meaning less than ten years for all the investment, and less than two years on a large portion of it, like TV work. Space development doesn't work like that, at least not in its current form. In order to launch a mission to Mars, a lot of money is needed up front, to develop rockets, habitats, rovers and to train crews. While I'm convinced that the Mars landing and consequent first steps on the Red Planet will be one of the most watched events in human history, I have my doubts as to whether or not the engineering and construction of a Mars habitat will be able to draw the audience for ad revenues to finance its building. Loans taken out on the promise that there will be a launch to Mars and that the revenue will be returned are unlikely, at least until something similar happens that a claim can be based on.

Property Rights

Another approach involves property rights. The argument goes that if the Space Treaty of 1967[79] were rescinded, or a signatory nation (such as the United States) were to withdraw from that treaty, then business interests would move into the solar system to prospect and mine. These businesses would land on the moon to mine helium-3 (discussed before as a useful fuel for fusion reactors someday), and would conquer the asteroids to gather minerals and metals for use on Earth. This prospecting and manufacturing wave would sweep a lot of people up in it, allowing them to fly into space to work and to live. Once again, I have no doubts that there are resources on the moon and asteroids that humankind should use and will need to use one day. In this case, however, I believe that humankind's inability to plan long term will kick in. The fact remains that investors give their money up with the expectation that they will be paid back with interest. The riskier the venture is and longer-term the investment, the more interest the investors will ask for. Currently, investments that are described as "risky" and "long-term" include: biotechnology research and development, household robotics, and new commercial satellite businesses. You'll note that none of these risky investments include launching humans into space on spacecraft never flown before, having them land on solar system bodies that have been visited few times or not at all by humans, then going outside to mine—and these are simplified versions of only a few steps involved in the process. When you leave mineral rights out of this, the picture gets even more difficult. Land on the moon is real estate, that's true, but what good is it? Unimproved lunar real estate cannot support human, animal, or plant life. It may serve as a platform for solar power panels, but that's an improvement, too. The only way the lunar surface can hold a use for humans and their counterparts is if a portion of it is enclosed and pressurized with air. Even then, prospects are dubious, because any plants growing in that pressurized area will require artificial light for the more than 330-hour lunar night. I think this idea gives a whole new meaning to the concept of risky investment. Yes, there is a man named Dennis Hope selling property on the moon, claiming that the Space Treaty has no bearing on individuals, only nations. Business is quite brisk, too, since the new initiative designed to take people back to the moon and on to Mars. The claims he's made have not met a serious court challenge yet, and something tells me that Mr. Hope will have difficulty defending them, either legally or physically, without the nations that he excludes in his plan right now.

Space Tourism

A third scenario related to private enterprise taking us to the stars involves space tourism. This argument follows that only increasing flight rate will lead to a large-scale cut in launch-to-orbit costs. Since there are no systems around today that require such a high flight rate, only space tourism, with the potential of needing many flights a week or even a day, can meet that need. Once space tourism takes hold, launch-to-orbit costs will plummet, driving the cost down of space exploration overall, and open the cosmos to humankind. I agree with the majority of this argument, but think that it's missing an important segment. It's true that today there are only about 50 flights into space each year around the world, and with a multitude of countries and even companies providing launch services, there aren't enough launches to make large-scale production and launch feasible. A high flight rate is the answer to this problem, but I think the idea of strapping paying customers into a craft that's flown a few times is a difficult sell. While it's true that the space shuttle flew with a crew on its first flight, that was a government program where rules are applied inconsistently at best. Any non-government agency who's providing a service to people will certainly have to face tougher regulation than NASA, and that will include flight certifying their craft. Under current guidelines, it's possible for the producer of a new commercial passenger aircraft to spend more money on certification than they spend on developing the aircraft, and this provides another obstacle for privately developed spacecraft. The United States Congress has considered legislation that would change the landscape for such spaceflight, however[80], and I eagerly await the outcome of their deliberations.

Benefactor

There's another possibility here, which can't be ruled out completely. There are people on this planet with a net worth equal to or greater than the estimated costs of some major space efforts. In the year 2000, Bill Gates of Microsoft fame (the richest man in the world) was reported to have a personal fortune of $55B. You can check his worth daily[81] if you like. That's more money than the cost of a late 1990s NASA estimate for a humans-to-Mars mission. If someone of that stature wanted to, they could fund a major space effort out of their own pocket. This type of space effort would combine many of the advantages of a fully government funded effort (central control, short timeline) with a commercial effort (lower cost, and with that, more application to exploration after the initial mission). To a smaller scale, this sort of thing is already happening. Jeff Bezos, the internet billionaire who maintained his title through the bursting of the internet bubble, founded a company called Blue Origin. Their plans are rather secretive, but it's rumored that they're developing a sub-orbital spacecraft for space tourism applications. Elon Musk, another wealthy internet entrepreneur, is bankrolling the SpaceX company, but not with the sole goal of seeing its rocket fly. Elon has built a business plan based on small satellite launches that he hopes to use to build credibility and parlay that into larger boosters. Given his track record at being successful in business, his ideas shouldn't be taken lightly.

Advantages

With the exception of the last scenario discussed of the wealthy benefactor, these possibilities for future space hold some advantages over a government-run program.

The cost pressures in a foreseeable entrepreneurial effort would be enormous. Whereas government programs are rarely cancelled after serious money is expended on the program (there is one notable exception, the Navy's A12, which was cancelled in 1990 by then-secretary of defense Dick Cheney) because canceling a program that's underway may makes people wonder why the program was approved in the first place, a project which builds up from the roots should have some more backbone to it. Part of the assumption here is that a mission beyond LEO taken on by a commercial effort wouldn't happen until there was some serious development in space infrastructure, making such a mission more of a step than a leap. Therefore, to make this model happen, the most important research to do right now would be ways in increasing flight rates, thereby decreasing costs per individual flight. Miniaturization and increased reliability of components, such as life-support systems and the like, would be another good focus area.

Disadvantages

One serious advantage of having a centralized program is centralized decision making, and a consortium of industry must plan carefully to have such control. When NASA ran the lunar program in the 1960s, there was no question as to who had the final say on any engineering or operations decision. Companies could offer their insight, but the leaders at NASA had the bigger picture, and could balance the needs of the mission with the needs of the various contractors. This usually led to the proper decision. In a consortium-based development, the lines will be less clear, and that will lead to difficulty. While it is possible for an industrial consortium (Airbus is a good example[82]) to produce impressive results, such collaboration takes an amount of planning and discussion that may be comparable to the planning required for the mission itself!

Without a central source of funding, there's also the possibility that one portion of the project could back out, putting the entire project in jeopardy. Let's say that 10 companies currently working in LEO decide to work together to put people back on the moon to stay. One company agrees to upgrade its launch vehicle to handle the larger payloads necessary. One company chooses to build a new communications infrastructure for the effort. Other companies take on the landing craft, surface suits, rovers, and all the other parts required for such a mission. Then, one of the companies runs into financial trouble and can't complete its portion of the project. Uh oh. Obviously, there are options such as finding another maker for the surface suit, but depending on when the original company backed out, this change could seriously impact the mission timeline, delaying launch for years. The delay could cause financial hardship for the other companies in the consortium, causing more backouts in a downward spiral.

My Opinion

I believe that an organic, multi-faceted, commercial development of space is a much sounder form of development than one almost exclusively developed by government agencies. I've heard the cries of some of space commercialization's detractors, claiming that the scientific purity of space will be lost and Starbucks bulletin boards will block the view of the stars as soon as businesses move out. These arguments don't hold a lot of water with me, not because I disbelieve them, but because 1) space is big, and 2) as long as we eventually get out there, I believe the overall benefit will eclipse any complaints about bulletin boards.

My belief in this approach is tempered, however, by the fact that I've not seen a realistic business plan that can take us there from where we are. I'm not privy to backroom dealings going on right now, but I've watched from a distance as many a space entrepreneur gave a dazzling presentation at a conference, only to declare bankruptcy in a matter of years. As the poster in Fox Mulder's *X-Files* office states, "I want to believe," but the facts surrounding a purely commercial effort taking us into space just don't bear it out yet for me. So, I spend time thinking of other ways it can happen.

SCENARIO 3: A HYBRID APPROACH

So, government won't provide a large-scale, focused effort without a concrete external threat and commercial ventures are likely to be spooked by risk and impending regulations. So, what's the answer? Here are some ideas of how government and commercial interests can work in concert to move us out of the cradle.

Preamble

In history, large risky endeavors are usually taken on by governments and then handed over to industry as risks lessen and revenues become obvious. Let's take a look at a couple examples.

In the early days of aviation, when aircraft were made of cloth and barnstormers traveled the nation offering thrills to state fairgoers with amazing feats of skill and bravery(?), an idea hatched that revolutionized air travel almost as much as the invention of the airplane itself. In 1911, the United States Postal Service gave a package to a pilot, who took it on board his plane, and flew it to a destination. This was the last time it was to happen for several years, but when the idea came up again, it caught on with a little more oomph. This time, in 1918, the postal service guaranteed to pay for flights between major mail hubs, whether or not there was mail on board the aircraft. This guarantee led many small companies to go into business providing these flights. Of course, there was likely arguments that such frivolous expense by the US taxpayer was a waste of money, and in the beginning, I probably would agree, but it led to something that we take so much for granted today that it's hard to imagine life without it…airline travel. In the process, it also created a crop of experienced pilots, those who flew these mail routes being employed by the mail providers. One of these pilots, a man named Charles Lindbergh, went on to fame as the first person to fly solo across the Atlantic Ocean.

As these routes became more established, and people knew that they were regularly occurring, there were times when the mail load increased beyond what an airplane of the day could carry, leading to larger and better aircraft. With larger aircraft, when there was room aboard, someone could travel along with the mail between two destinations, and pay for the ride. Eventually, enough people saw that this kind of travel wasn't killing thousands of passengers each year that they decided they could travel on aircraft as well, so people started asking for more flights. Soon after that, aircraft were built with dedicated passenger space, the first commercial passenger aircraft. Pretty cool, eh? But this type of thing could never work for space, right? Well, actually, it already has.

Any time you turn on your television, you are using an asset in space. Don't believe me? Let's take a look. The obvious sights to see are houses with satellite dishes on their roofs or hidden somewhere inconspicuous so that the homeowner's association doesn't write the owner up. These homes are taking satellite data beamed directly from space into their house and putting it on TV. If you have cable, it's true that you don't take your signals directly from a satellite, but the cable company that sends TV to you has satellites in their links somewhere. If you're one of those people who doesn't believe they should pay for television, but still watch the evening news with live broadcasts from around the world, once again, you're taking advantage of satellite communications.

The Earth is literally surrounded by hundreds of communications satellites, most of which are in a belt call geostationary orbit. There is an excellent illustration of this on the web if you're interested[83].

Geostationary orbit means that the satellites always appear at the same spot in the sky if you were to look at them. More importantly, it means that an antenna pointed at them doesn't need to change position. The satellite communications industry dwarfs all other portions of space spending every year, and there's actually a very small fraction of the data being transmitted that is owned by governments. How did all this start?

It turns out that President Kennedy was interested in other achievements in space besides putting a man on the moon. In 1962, he signed the Satellite Communications Act into law. This act created the Communications Satellite Corporation (known as COMSAT) whose job it was to commercialize satellite communications. This is one case where the space administration (NASA) demonstrated that a technology was feasible by launching a communications satellite, then backed away from the COMSAT corporation and let them work. Soon after COMSAT was formed, INTELSAT, an international organization dedicated to global satellite communications, came into being, and COMSAT was the sole US signatory on the treaty. Over time, as the communications arena changed, and communications satellites were overtaken by underwater fiber-optic cables connecting the continents in data rate, COMSAT became less of a player. Early pricing rules that gave the corporation an advantage prevented it from branching into new arenas. Meanwhile, other companies were founded that used satellites for communications. The most successful to this point, direct-broadcast TV, required huge satellites of enormous power, but since COMSAT largely demonstrated the technology through the years, investors could see that the new requirements were not out of the question.

Mechanism

So, pulling these lessons together, how can a hybrid government/commercial effort work? Here, my theories are going to sound very similar to the Space Frontier Foundation, so I want to thank them in advance for providing the seeds of many of the ideas.

First off, the US Government must stop developing and operating space launch vehicles. Any time the government puts out a contract for the designing, building, and operation of a space vehicle, the system works something like this:

1. The agency awards a series of study contracts. Companies use these funds to research different designs of a spacecraft.
2. A series of "downselects" decreases the number of contractors as dollar values go up for more study contracts. Eventually, the downselects reach the point where a final request is put out for one contractor to build the final product, and the remaining contractors respond to this request. Each response is called a proposal.
3. A group of career bureaucrats (usually, they are very technically adept) gets together in a room and weighs the merit of each proposal.
4. Political considerations (such as which congressional district the work will be done in and who won the last contract) are taken into account.

5. The contract is awarded, sometimes counter to the merits weighed in item 3 or 4
6. The contractor and the government agency get together to discuss what was really desired, usually in an entirely different money situation than was proposed.
7. The redesigned vehicle goes through a series of design reviews, where increasing details are added to the plan.
8. The vehicle is redesigned again, based on new budgetary guidance from Congress.
9. The contractor starts bending metal, actually building the device.
10. Another redesign takes place.
11. Costs and tensions between the contracting agency and the builder rise, partially due to the number of changes required, but also due to the number of things being done for the first time. It is difficult to estimate how much things will cost when they've never been done before.
12. Eventually, the device is built and flies. Nearly all tensions are forgotten, and awards are given out for the technical marvel that is the new device.

I can't fault this system for eventually reaching a goal—it got us to the moon and built the space shuttle (as well as aircraft carriers, airplanes, and most everything else built for the US Government). I will go out on a limb, however, and say that there are development systems that carry out system developments at a lower overall cost. This is one area of cost savings that should be considered.

So what I picture is a government program where money is paid when cargo is delivered to orbit. Essentially, the government procurement process becomes one step. This is close to the way that people deal with shipping on the planet, because if you want to send a package to your aunt in Moose Lake, you don't have to create an entire package delivery system. Someone else has already done that, and you just have to pay for their services. Because this type of delivery has been done for years, you don't pay on delivery (when your aunt receives your package), you pay when you drop the package off.

While it's true that this approach turns the government procurement system upside down, it guarantees that money isn't thrown down rat holes of dead-end research. If you've followed space development over the last years, the field is littered with partially built programs that cost taxpayers billions of dollars overall while producing zero flight time. Examples include the

National Aerospace Plane (NASP), National Launch System, X-33, and X-34.

Secondly, some form of stable regulatory environment must be created to allow a corporate start-up to know the rules it will operate under. This sounds simple in principle, but because travel to orbit has so many facets that are unexplored at this time, the regulators are at a loss of what to do. When regulators get lost, they tend to over-regulate. I have a hard time blaming them, because if something terrible happens with a new space system that causes injury or death to a civilian on the ground (up until now, the US space program has a perfect record of zero civilian casualties), the regulator who championed a loose regulatory environment will be called in front of Congress to explain why they didn't think regulation was necessary. While most bureaucrats would be saddened by the loss of life, they'd much rather go through major dental work without Novocain than face a Congressional inquiry. So, the road that is taken is the easy road for the bureaucrat, but the difficult, and maybe impossible, road for the space entrepreneur. You'll note that I didn't say that the bureaucrat's job would be in jeopardy. I say that because our nation's government went through the September 11, 2001, attacks without firing anyone, so it's unlikely that the loss of someone in a space accident would cause any firing.

One thing that makes this regulatory environment so difficult is the number of ways that have been proposed to go into space, along with the number of flight profiles proposed. So far, ballistic (rocket, or modified rocket in the form of the shuttle) has been proven to work well enough for continuation. This has all proceeded either under government control, or with close government supervision. The new players in the game have other ideas, however, to match their new goals in some cases. Burt Rutan has a carrier aircraft that carries a smaller craft to altitude, then drops the smaller craft so it can fly to its own high altitude. Is this a spaceship? By classic definition of space beginning at 50 miles up, it is. Theoretically, anyone who crosses that line should get an astronaut pin, and the FAA revealed a commercial version of just such a pin, awarding it to the pilots of X-Prize winning *SpaceShipOne*. Pins are symbolic, but who's going to certify that this craft is safe to carry paying customers? There are other examples.

Rocketplane Limited, formerly Pioneer Rocketplane ,has worked on a craft called Pathfinder for some time. Their concept has Pathfinder take off under its own power from a runway. It flies to normal aircraft altitudes, and rendezvous with a tanker aircraft. The tanker craft transfers liquid oxygen to the rocketplane to fill its tanks. Once full, the rocketplane separates from the tanker, ignites a rocket engine, and flies to altitude. How should this be classified? It takes off like an airplane, takes on propellant, flies very high, then lands like an airplane. Right now, in-flight refueling is something left to military planes, and they transfer fuel—Pioneer Rocketplane's design transfers liquid oxygen, which is much different. How should this be licensed? This conundrum came to light recently and there were some congressional hearings held on it. Recently released legislation (again, HR 3752) seeks to clarify everyone's roles in a commercial space flight, to the point of creating a passenger class that is not the same as passengers on board commercial airliners. This definitely has the makings of a step in the right direction.

NASA should be more concerned with research and standards than with operational hardware. The predecessor to NASA, NACA (the National Advisory Council on Aeronautics) played this role nicely. They funded research into early instruments, and developed a standard series of wing types that they fully characterized. That data proved tremendously valuable for aircraft designers. While it is more difficult to get a vehicle into space than it was to get one off the ground in the early 1900s, NASA could still serve an outstanding role answering critical questions for a burgeoning space industry. Here are some examples:

1. What's the best shape for a lifting body that can return people from orbit?
2. What type of engines provide the best mix of thrust and overall performance?
3. What is the best way to use Mars' atmosphere to slow down on approach to the planet?

To some extent, NASA does this research now, but not to the point of finding the best design. They may demonstrate that an ion engine powered by concentrator solar cells will work (this was done on the Deep Space 1 mission), but in the race for press hype of the mission, the ion engine used is not compared to others that could have been chosen.

Another role that some descendant of NASA could take would be similar to that of the Coast Guard or Civil Air Patrol, in the ideas of space rescue. Rescue in any media has typically been in the pervue of the government, so an extension into space is a logical road to take. After setting certain standards that a spacecraft must meet in order to fly from the US, including minimum supply levels in case of trouble, approach methods and the like, the government could have rescue craft on standby with crews ready to go in case of trouble. Once word of a problem came in from orbit, the crews would await a proper launch time, then fly up and rendezvous with the stricken craft. Using established procedures, the rescuers would approach the damaged ship, transfer the crew to their rescue vessel, and bring them home safely.

Advantages

If the heavy lifting of carrying propellants and eventually supplies are handed over to commercial interests, NASA and other space agencies would then be free to devote more of their resources towards exploration. This is where orbiting depots or some other idea comes into play, to bring about the unheard of flight rates and the accompanying reductions in cost and increases in reliability.

The way this allows space agencies to focus on other missions is by removing the burden of having to worry about getting their craft to orbit. If many commercial providers are available and make routine flights into orbit, success rates will improve dramatically, likely approaching those of airliners. The entire corps of engineers that validate launch vehicle processing today will be free to pursue other interests. Hopefully, they would stay within NASA and carry out further exploration or technology development.

As costs drop, the spaceflight industry could synergistically feed into the space tourism market that is often hyped but difficult to envision without some new legislation. Once a market exists, other entrepreneurs will find it easier to get money to try their new method of achieving orbit, and some of them will actually have a good idea that will revolutionize the industry again.

Disadvantages

I've been unable to think of any disadvantages of this system, but I'm sure that others will suggest some if word spreads about the idea.

My Opinion

In case it hasn't been obvious in my writings, I believe this is the way to go. This approach of allowing governments to handle long-term efforts while corporate entities take on the short-term has proven itself over and over in the past, and I believe it will work again.

Governments are currently "doing" most of space right now, and an entrenched bureaucracy is one of the toughest infestations to fight. I've witnessed government employees' angst over the simple changing of job titles and reporting officials, even though there was no chance of anyone actually losing a job. People who have powerful friends in government will fight any perceived threat to the space establishment.

So, things have to happen subtly. If commercial interests are allowed to handle lifting mass to orbit, mass that isn't important in individual portions, but important overall, they'll be able to cut their teeth and build capability. In the meantime, government agencies will be able to build the "high tech" portion of the space missions, keeping the entrenched interests busy doing what they love to do. With each side playing to its strengths, it all may happen.

Chapter Nine

Yes, but What Should Happen Now?

Through the course of writing this book and listening in on the "national debate" about the space program, I've come up with the following recommendations. Other people in various venues have proposed many of them. I've taken what I think to be the best recommendations out there, combined them with some of my own work, and listed them here. They have varying realism in being followed, but should provide for some excellent discussion. I'll likely catch some flack for parroting many of the goals set forth by NASA's new direction, but I can honestly say that these were written before the January 14th speech by President Bush. When asked to prove it, I'll show my e-mail records to my manuscript backup site that shows a January 5th e-mail including these items. Besides, I'm a big believer in there being no such thing as bad controversy, if a lot of people get talking....

Set a Goal: Mars, When Private Suppliers can Lift More Than Half of the Payload to Low Earth Orbit

It's been said more than once that Mars has been the next logical goal for human exploration. The primary arguments that I've heard against it is that a mandate to go to Mars right away will have too much in common with the mandate to send humans to the moon in the 1960s. In the Apollo era, we built the machines that took a small number of us to the moon, accomplished the task, and then abandoned it just as we started sending scientists to fully characterize it. This is a valid concern, and it's the reason that I don't recommend setting some sort of short timeline for the task.

In the Apollo moon race, America won by choosing an exploration method that minimized the amount of mass necessary to accomplish the goal. The lunar orbit rendezvous approach achieved the minimum mass possible at the time. Yes, there was minor dissention in the decision of using Apollo capsules when the two-person *Gemini* capsules could have accomplished the same goal in a shorter time and spending less money[84]. Once the Apollo system (including the Command/Service Module [CSM] and Lunar Module [LM]) was settled upon, a quick and dirty approach to getting a large amount of mass into orbit in one shot (the Saturn V moon rocket), was built to push the necessary hardware to the moon. Other options were considered, including launching one craft, approximately the size of the Atlas rocket to the moon with three people aboard. This approach was called direct ascent, but it was discarded because a rocket capable of lifting it off the Earth would have made the Saturn V look small and would have taken too long to develop. Earth Orbit Rendezvous, where smaller rockets launch parts of the lunar exploration system into orbit, which then join together for the voyage, was also discounted. The official reason was that it would take too long to build up the infrastructure in orbit to do so, but I suspect that part of the reason was that Werner Von Braun wanted to build a really big rocket.

So, to avoid the "crash" approach taken by Apollo, this mandate for Mars gives NASA and the other space agencies of the world the opportunity to build the hardware for Mars exploration, while at the same time, tasking them to work on low-cost access to space. The low-cost portion of the mission is possible because when a Mars mission (built to fly using chemical propellants) is in low Earth orbit, its mass is made up of approximately 70 percent propellants. Propellants can be carried to orbit relatively easily, either using the orbital supply depot described earlier or by carrying the fuel (hydrogen) and oxidizer (oxygen) separately. By contracting to have these propellants delivered on orbit, space agencies can then focus on building the necessary hardware for exploration, not propellant storage (as the tanks holding hydrogen and oxygen specialize in, even though they supply rocket engines and make the mission possible). With a burgeoning commercial launch industry coming together for delivering propellants to orbit, combined with a new round of development for spacecraft to go beyond LEO, a true renaissance in space travel may be upon us.

EXPAND SPACEGUARD

The budgets here are very small compared to the others discussed, but I wanted to mention it. The Spaceguard survey, a loose conglomeration of researchers and agencies around the world, is cataloging near Earth objects (NEOs) that have the potential to impact Earth. They are doing this on a budget of mere tens of millions of dollars. Government agencies should increase their contributions to this effort, focusing on improving the hardware doing the survey instead of the number of people involved.

BUILD A LOW-COST UTILITY VEHICLE

A new space vehicle is necessary to take on new challenges in human space flight. I am biased towards a capsule design, because I believe that it is more flexible in serving multiple roles. This capsule need not be reusable, because it's likely that such reusability will raise the price significantly. Not making the craft reusable may actually come to decrease costs over time, as a production line for the capsules could remain open, and over time, new capsules could be upgraded. Something similar happened during the Apollo program, as new technology and techniques were applied to make the lunar module more capable, allowing longer missions to the lunar surface. The key in this approach will be production rate, which will depend on how often such craft are to fly.

The utility vehicle, once built, can serve a variety of roles. As described earlier, this craft could take travelers to the space station, and serve as a long-term escape capsule for the outpost. One of the capsules could also sit on the ground, ready to go on short notice as a rescue vehicle for any craft having difficulties in orbit. As humans move further out from Earth, on journeys to the moon, the asteroids, and eventually Mars, they will need to reenter Earth's atmosphere and land on the surface, and this same utility vehicle could be used for those missions as well. By giving the return-to-Earth role to the utility vehicle, the craft that travel away from Earth and support the crew for the majority of the time are not constrained by aerodynamics, and therefore can truly be space vehicles.

Open Orbital Supply Depots

As described earlier, orbital supply depots provide a practical solution for enabling frequent commercial access to low Earth orbit. By offering money for the delivery of bulk supplies (water, for example) to orbit, agencies have the capability to unleash the power of free enterprise on the vexing problem of cheap rides into LEO. As these supply depots grow in size and capability, vehicles that are traveling beyond LEO will be able to take on fuel and oxidizer for their rockets and water and oxygen for their crews before pressing on to their final destinations. In the beginning, it's likely that those destinations would be high Earth orbit for equipment checkout, followed by lunar and asteroid tests, with the ultimate destination being Mars. For supply depot operations on this scale, it's likely that there will have to be multiple depots in multiple orbits, with at least one serving each of the major launch sites, plus any others that crop up because of the new open competition for launch capabilities.

Complete the Space Station to Meet International Treaties, Then Turn it Over to an Authority

The plan for building the space station has been published in various forms over the years, and it's readily available for anyone who's interested. Plans for operating the station after completion are less forthcoming. Another wrinkle comes into play that NASA, the banner agency building the outpost, is not charged with operations, but instead with research and development. An example of this division can be found in Earth-sensing spacecraft. The weather satellites that provide up-to-the-minute images of the Earth are operational in nature. Though many pictures of hurricanes and the like credit NASA for the image, the National Oceanographic and Atmospheric Administration (NOAA) is actually the agency charged with operating the satellites. NASA plays an important role in purchasing the satellites for the government, and does some excellent work inserting color into the images, but once those satellites are in orbit and operating, they become a NOAA asset. NOAA is charged with delivering the raw satellite data.

The space shuttle has gotten around this because it's been described as a research and development vehicle by several outside agencies, often in contrast to NASA's own efforts to try and declare the STS operational. The

Gehman Commission concurred with the idea that the shuttle is not operational and that effort to try and treat it that way is misguided.

A space station, with construction complete and regularly scheduled flights to it for crew rotation and cargo delivery, will have more trouble holding on to a research and development moniker. Granted, research and development will go on within and around it, along with (hopefully) a lot of hardware demonstration for use beyond LEO, but the day-to-day maintenance, scheduling, and the like will be an operational activity. At this point, it falls outside of NASA's expertise.

In a series of congressional meetings in 2000-1, The Space Frontier Foundation proposed that, once construction of the station was complete, it should be turned over to the space equivalent of a port authority. Port authorities, as they exist here in the US, are a unique blend of governmental and commercial interests. When necessary, they have the clout of a government agency, but are not completely bound by the procurement laws that tend to slow down a full government bureaucracy. The exact details of each port authority's power are spelled out in its charter, and the first space authority would be quite an exercise in international law. Once established, the authority would have the ability to contract for station resupply, they'd also have more freedom to look at other options for the station, including, perhaps, adding large pressurized volumes to the outpost for increasing short-term occupant capability on board. The idea of space tourists visiting the station would be less of a taboo subject than it's been in the past with NASA running the show. All this could take place with less public funding, as the authority could have more freedom in raising funds through some creative ways, such as filming the station or views from it for profit, as well as carrying out commercial research on board.

NASA would still play a role in the station as long as they were interested. First off, much of the research on new propulsion systems and life-support systems would likely be sponsored by the agency. Secondly, NASA currently holds the majority of information related to the station: its capabilities, liabilities and future expansion prospects. They would have to be involved with any plans to expand the station, and as long as the US Government maintains some sort of launch capability to the station, it's very likely that NASA will be a major player. Jobs that are within NASA today that deal with operations of the station would simply transfer to the authority. Over time, it's possible that the authority would find better, more efficient ways of doing things, and as they're implemented the number of people working on the station from the ground could drop, but short-term employment fears, often a primary consideration in consolidations, should be minimal.

Tom Hill

Discount the Space Station from Future Flights Beyond LEO (Unless...)

The space station was originally envisioned during the waning days of the cold war as the United States' answer to the Soviets' continued presence in space using their own space stations. As the station was designed and redesigned, its purpose changed to that of keeping Russian scientists and engineers busy as the former Soviet Union crumbled. To do that, its orbit was changed from one easily reached from the US to one that was harder to reach from the US but was feasible to reach from Russia. Doing that, the powers that be put the space station in an awkward location for flights beyond LEO. It takes more fuel to fly one kilogram of payload into the space station orbit from the United States or from the European launch site in South America than it takes to put the same kilo into their optimum orbit. To be fair, it takes much more fuel for a Russian rocket to launch a kilogram of payload from their launch sites into orbits optimized for the US and European Union. When flying beyond LEO, any kilogram that's harder to launch to LEO is even tougher to move beyond, and that cuts significantly into the amount of payload that can be sent to the moon, the asteroids, or Mars. Unfortunately, because so much money has been spent on the station, I find it hard to believe that today's space agencies will "abandon" it by not using it as a stopover for flights on their way to distant points. To carry out such a stopover would be a tremendous waste of fuel and launch capability, and should not take place, unless something pretty radical happens. In current NASA goals as stated by President Bush, it appears as though the station will not be used as a stopover for missions beyond LEO.

In an internet post, moving the space station from its current orbit after completion was proposed. While this would be a tremendous effort, requiring ion engines thrusting over months or years to accomplish, there is nothing in the rules of orbital mechanics that says it can't be done. Of course, there may be some constraint in the way the station was built that would prevent the idea from working, although in a simple "hey, would this work?" e-mail to a buddy of mine in the space station program, only the good things about it happening came back. I get the impression that the question didn't get a lot of serious consideration. Spacecraft are usually designed specifically for the orbit they will inhabit, and some pretty important items (solar array pointing, sun shades over attitude sensors, things like that) are dependent on being in the proper orbit. Moving the station would also end Russian flights to the station from their original launch sites, although there is talk of their building a launch capability from the European launch facility in South America[85]. If the station were to be moved, its new location would dictate the best launch site for missions beyond LEO, and then, flights from that launch site would suffer less of a detriment (though no real benefit, unless an orbiting supply depot were co-located with the station) for a stopover on their way beyond LEO.

AFTER STATION COMPLETION, END CREWED SHUTTLE FLIGHTS

The space shuttle was designed as the first of a three-step plan for taking humans to Mars. In this step, the cost of flying into space would be lowered so that more complex missions would take less of the federal budget and become more feasible. As discussed previously, the shuttle instead became a dead end without a place to go. When the space station project was approved, step two in the earlier plan to take us to Mars, the shuttle had a place to go, but first the place had to be built. Between space station approval and today, two space shuttle missions ended in disaster, killing 14 astronauts and causing separate reexaminations of the space program. While some feared that the loss of a second shuttle would cause the end of the space program, February 1st, 2003, didn't appear to do that, as within days of the accident, the President reaffirmed US commitment to space, strengthened by the policy speech of January 14th, 2004.

Because the station is a project tailor made for the space shuttle, the individual pieces of the orbital outpost are designed to fly on the space shuttle. The pieces launched by the US require assembly in orbit, and don't have the autonomous assembly capability that their Russian counterparts have. So, because of decisions made early in the space station program, the shuttle must remain active until the station is completed in some form. The current plan has the station being completed to the US Core Complete configuration, which would hold crew levels at three, and would provide space for all the international participants to place their own components on the station. This configuration would meet the letter of the treaties relating to US involvement in the project, although the lack of personnel on board (originally planned to be 7), would impact the number of international researchers on board to use the internationally built facilities on board.

After reaching that point, a recommendation by the Columbia Accident Investigation Board to separate people and cargo should be implemented. This recommendation was confirmed by Admiral Gehman, head of the board, in Senate testimony on October 29, 2003. The space shuttle was designed to carry crew and cargo because it was meant as the be-all, end-all space transportation system (hence its name). The CAIB concluded that trying to meet all those requirements created an inherently fragile vehicle, more fragile and susceptible to damage from cold (*Challenger*) or debris (*Columbia*) than a vehicle would be that was designed for either crew use or cargo use exclusively.

The space station will have to carry on using one or two forms of transportation for the crew including the *Soyuz* vehicle from Russia and, hopefully, the utility vehicle described above. Other types of transport will haul up regular cargo, as the automated Russian *Progress* spacecraft does today, but it will be joined by the European Automated Transfer Vehicle (ATV), and it's likely that another delivery method will be necessary for large items, such as space station additions.

So with the station as complete as it will get, some people say that the space shuttle orbiters should be upgraded to fly autonomously. This is a bad idea. While it is possible to do, the Russians flew their space shuttle orbiter *Buran* (*Snowstorm*) once, uncrewed around the globe in 1988, just because something can be done doesn't mean that the idea makes sense. We'll go into more detail later in the next section, but the full space shuttle stack, solid rocket boosters, external tank, and orbiter, have the ability to put 110 tonnes into orbit near the ISS (120 tonnes to other orbits). The problem comes in that close to 100 tonnes of this mass is made up of the orbiter, with only 12 tonnes left for cargo. It's true that the removal of life-support systems, currently pretty important to the people flying on board the craft, will increase the orbiter's payload capability some, but this increase is unremarkable. The bottom line is that converting the orbiter to fly autonomously would turn the space shuttle system from a fragile launcher of people and cargo, capable of flying only a few times a year, into a fragile launcher of cargo, flying the same number of times a year. There is another option, however….

CONSIDER REMOVING THE ORBITER AND CONVERTING THE SHUTTLE STACK TO CARGO CAPABILITY

The space shuttle system, at liftoff, generates approximately 6 million pounds of thrust. The Saturn V rocket, which took humans to the moon, generated 7.5 million pounds of thrust. Despite the shuttle's having 4/5 the power of the Saturn, the shuttle can only carry 12 tonnes of cargo into the ISS orbit. The Saturn V could carry over 100. As discussed above, this is because the orbiter also has to be carried into orbit. If the orbiter were removed, and a cargo pod with three engines as powerful as the orbiter's are today and some maneuvering capability were mounted on the external tank instead, this newly configured system would be able to carry at least 80 tonnes into the same orbit. Other modifications, like adding more engines and/or moving the payload from the side of the vehicle to the top and adding another stage, would create vehicles capable of launching even more payload into orbit, in the realm of 120 tonnes. Now, we're talking about lifting some serious mass into low Earth orbit, and, as quoted before, according to the great science fiction author Robert Heinlein, once you get into LEO, you're halfway to anywhere.

This vehicle, in many configurations, has been proposed before. The first incarnation, the Shuttle-C, describes the first idea of adding a cargo pod with engines, though concept diagrams show two engines on the back of the cargo pod, which would decrease its payload capability. Other names tossed around are the Shuttle-Z, proposed by a NASA subset, and Ares, proposed by Robert Zubrin. The Ares takes the widest departure from the current shuttle system, and therefore is the most expensive option, but also carries the greatest payoff as far as launch capability into orbit. Robert Zubrin called for the shuttle-stack-to-cargo-only conversion most recently at the 2003 Mars Society Convention, followed up by several opinion columns in newspapers around the nation.

Such a conversion, instead of a completely new design, has advantages. It takes advantage of existing infrastructure and personnel experience. Infrastructure like that can't be overlooked, and anyone who's visited Kennedy Space Center knows why. It's HUGE and EXPENSIVE. Any new system would require new infrastructure, or serious modifications to the current infrastructure. This needs to be kept in mind for any alterations to the shuttle stack as well, because large alterations to the stack could involve large alterations to the infrastructure. Also, remember that when flight rates are low, the most expensive portion of any activity is personnel. Now, while an idealistic version of this plan has missions flying to the moon, the asteroids, and Mars along with checkout flights for all the hardware, it's unlikely that the flight rate of a heavy booster to support that activity would rise above 10 flights a year. Therefore, it's likely that ground personnel would be a significant cost driver of such a system.

While it's true that this recommendation, if followed, would provide a heavy-lift capability for the United States in relatively short order, I cannot whole-heartedly recommend it (hence the "consider" dodge at the beginning of the topic). The shuttle, as discussed earlier, is an old system. Because it is maintained with such care, it is not fair to classify it with household items that are its age. Not many people would still use a VCR first used in 1981, but there are many aircraft from that era still flying every day. Therefore, it's feasible for the shuttle to keep flying, and for a modified stack to carry huge amounts of payload into LEO for destinations beyond. The problems come in because the shuttle can still fly, but much of the support infrastructure used to support it is past its prime. This goes beyond rust-holes in the Vehicle Assembly Building, used to assemble the shuttle before it moves out to the launch pad. Here, we're talking about industrial parts: computer processors, actuators, and the like that make the shuttle system go. The orbiter has undergone some tremendous upgrades over the years, as has the external tank, but upgrades to the solid boosters have been lacking. In one highly publicized case, parts actually had to be taken from a museum piece to allow a flight pair of solids to fly[86]. These problems are not show-stoppers, in that a few upgrades to solid rocket booster systems should enable a lower-cost system, but upgrades cost money and cut into the savings originally advertised. In another area, if flight rates eventually are required beyond anything imagined today, this modified cargo launcher would again have to change with the times, costing more money.

There's also a more sensitive issue, in that a government-owned heavy-lift vehicle is likely to prevent private interests from developing a heavy-lift launch vehicle. This could easily (though perhaps not cheaply) be remedied by paying for deliveries of large payloads, similar to the smaller payload system used as part of supply depots, but that's a discussion for another day.

PROOF-TEST MARS HARDWARE IN LOW EARTH ORBIT AND HIGH EARTH ORBIT, THEN ON ASTEROID AND LUNAR MISSIONS

One of the drawbacks of a Mars mission discussed earlier was their length, going months or years without materiel support from the home planet. Without exotic technology propulsion, beyond most of what's even on the drawing boards today, this concern will not vanish. It can be soothed over time, however, by showing people that it is possible for Mars-mission-built hardware to keep a crew alive for increasing periods of time. This will likely not be a smooth process. Early attempts at long-duration spaceflight in the 1960s (14 days was the goal back then) were delayed by equipment development problems, and it's likely that the same type of issues will arise as we try to build hardware that will support our travelers for months on end without replacement. That's OK, or at least it should be, although taking setbacks in stride hasn't been a strength of national space agencies, their government budgeters, or the media agencies that cover them.

With the hardware proven in LEO and high Earth orbit, exploration can begin again for new lands. Returns to the moon will now be possible, and it's likely that the on-orbit rocket propellant industry will be willing and able to sell the necessary supplies for the trip. Asteroid exploration can proceed in parallel, as launch windows open to our near neighbors orbiting the sun, further easing fears that such things are too difficult for us to handle.

GO TO MARS WHEN THE HARDWARE IS READY AND PRIVATE ENTERPRISE CAN MEET THE DEMAND

If this plan is implemented, sometime in the not-too-distant future, there'll be a huge propellant production capability in LEO, serviced by spacecraft flying from Earth's surface at unprecedented rates. Propellant production also leads to propellant, and vast stores of material will be in place for use by anyone traveling to orbit that is willing to pay for it. The flight rates discussed earlier will cut the cost of flight to orbit, and a number of industries, some imagined, some not, will be in place or on their way to take advantage of the new going rate for flight beyond Earth's atmosphere. People will have returned to the Moon, but also traveled to an asteroid or two, all the while testing power, life support, waste recycling, and other systems required to make the journeys easier and lighter. People will have spent months in high Earth orbit, verifying radiation shielding properties and documenting the impact that reduced gravity has on the human body. The public, growing used to 2-3 launches a day from Earth, doesn't give much of a thought to the occasional loss of a cargo vehicle, with statistical trends showing that such events are becoming rare. Human flights are aborted at times, but crew capsules pull away from their launch craft the crew walk away from the incident. We will be ready.

One day, a heavy-lift launch vehicle will take off from its assigned launch pad. Even though it looks like many of the others that take off from Earth all the time, this one is different. It carries the hardware to take people to Mars. The land around this launch site is covered in people who want to see a part of history, and they'll stay in town for the planned launch of the crew two days later. The cargo from the heavy vehicle will travel to orbit and dock with an orbiting supply depot. There, it'll take on its supply of hydrogen, oxygen, and water. The crew will dock with the vehicle, transfer their personal effects aboard, say goodbye to the vehicle that brought them up, separate from the supply depot, then rev up the engines on board their spacecraft. We'll be on the move again.

Idealistic? Yes, but it's possible.

As far as timelines go, I believe this sort of thing is doable in 20 years. Others call for 10, and while that length of time is extremely tight, but feasible for a humans-to-Mars program, to me, it too closely parallels the Apollo landings. Still others call for no mission to Mars until the Earth-Moon system is completely exploited through commercial means. Once again, this is feasible, but it ignores the realities that we're in today, because space is currently dominated by governmental agencies. Debate the wisdom of our current situation all you want, but it is our current situation. There needs to be some sort of transition plan where commercial entities can start showing their worth. In this plan, the existing government agencies do the long-term research and development (their strength) and commercial entities do the repeated lifting and processing (their strength).

Chapter Ten

The New Space Policy of the United States

On January 14, 2004, President Bush traveled a few blocks from the White House to NASA headquarters. His speech was anticipated, expected as a response to the Columbia Accident Investigation Board's call for a challenging vision for the United States space effort. Many thought that the new policy would be announced at the 100th anniversary of flight celebration in North Carolina, but they were disappointed when no announcement came. The week before the speech, word leaked out through United Press International about its substance. A transcript can be found on the White House web page[87], along with the accompanying policy.

In his talk to invited guests, President Bush spoke about the space program and how it was important to America. He stated that the combination of technological innovation, educational inspiration, and national spirit that the program carried made it that way. He then went on to spell out the next steps for the space program.

The Goals and Timelines

President Bush's speech set the direction of NASA. In it, he specified that the space station is to be completed in accordance with the international treaties that the US has agreed to. In order to do this, the space shuttle must be returned to flight, which it's currently scheduled to in the year 2005. Once the space shuttle orbiters start to fly again, they will be devoted to assembling the space station. There are currently about 25 more construction flights required for the space shuttle in order to finish its job of carrying large parts to the station, and assuming a flight rate of about four per year, the last part should be carried to its destination at the end of 2010. Upon completion of the station, the space shuttle system will be retired.

The space station will have its focus shifted exclusively to preparing for human spaceflight beyond low Earth orbit.

In the meantime, NASA has been tasked to create a new spacecraft, currently called the Crew Exploration Vehicle. This vehicle will serve in the roles that were once assigned to the Orbital Space Plane, which was cancelled as part of the new policy, but will also serve the needs of deep-space exploration. The CEV is to be designed in a modular fashion, so that it can be "suped-up" for missions where additional capabilities are necessary, but has basic flight modes that can be used for missions that don't require the additional mass. Examples of one company's ideas for such a vehicle can be found on the web[88]. According to the President's address, the CEV is to start testing in 2008, and fly in a crewed fashion by 2014.

While manned spaceflight presses for the completion of the station at the same time as it's building a new vehicle, a new robotic effort is to start pressing its way toward the Moon and Mars, with the overriding goal of preparing for human exploration. The focus is to be demonstration of enabling technologies that crewed spacecraft will eventually be able to use.

After completion of the space station and retirement of the shuttle, effort will switch to lunar exploration as a testing ground for travel beyond. In his speech, the President said that crewed flights to the moon could begin as soon as 2015, but no later than 2020. Time on the moon will be spent learning to use its resources, and preparing to take the next step. Though it's been widely reported that President Bush's speech called for a trip to Mars by 2030, the transcript of his talk does not reflect that date. The Mars goal is made clear, though.

THE BUDGET

As with any newly proposed project, the first question is "How will we pay for it?" but the speech and ensuing policy dealt with that question. For the first steps, taking place between now and the completion of the space station NASA has been given a budget increase of $1B in the 2005 budget request. The President also spoke of a $1B increase in the NASA budget, taking place over the next 5 years, and his budget request reflects that. This money is to be supplemented by using one of the new policy's more controversial measures: canceling programs that do not support the new vision. At this writing, it is unknown how much additional funding the cancellation of those programs will produce.

Once the space station is complete and the shuttle retired, the combined budget of both programs is to be moved over into the new initiative. This amounts to about $6B a year. So assuming that the space station is completed in 2011, that the realignment of NASA programs produces $2B in savings, and that current increases are approved, there's a total of approximately $100B dollars devoted to this program between now and 2020. (That's $3B/year for the next 7 years through 2011, followed by $9B/year for the years from 2012-2020.)

Critics immediately attacked the plan on the basis that it would cost way too much to be feasible. While it's true that estimation of spaceflight costs are notoriously difficult to make and NASA has gotten into trouble over spiraling costs of the space station and other projects, Sean O'Keefe, the current administrator, fundamentally changed the accounting system of the agency. This should alleviate some of the problem. Another problem with the charges is that they're made based on the cost of the Apollo program, where the United States mobilized to achieve the landing-on-the-moon goal in a matter of less than ten years, and the response to the 1989 Space Exploration Initiative. As mentioned before, NASA's response to the Space Exploration Initiative was to bring out every idea that was ever proposed for a mission to Mars and include it as part of the package. This led to an unworkable plan. If, however, fiscal sanity is applied, current resources are used to their best benefit, and new development limited to that which is absolutely necessary, I believe that this new challenge for America's space program can be met. That's where the Aldridge Commission comes in.

THE PRESIDENT'S COMMISSION ON MOON, MARS, AND BEYOND

One portion of the new space policy, which differed slightly from efforts like it in the past, was the creation of a commission to oversee and provide recommendations to NASA in its execution of the policy. As part of the announcement, Pete Aldridge, former secretary of the Air Force and president of the Aerospace Corporation (in the spirit of full disclosure, he was my boss at one time) to head the commission. President Bush gave the commission 120 days to form its opinions on how the space policy was to be carried out.

Mr. Aldridge made it very clear at the first public hearing for the commission that their charter is to make recommendation as to how the policy is to be implemented, not to recommend changes to the policy. There were a number of public hearings held over the early months of 2004, after which they submitted their report, and disbanded 60 days after that. The commission has created a website[89], and now, with their report submitted, the website links surfers to a NASA website that thanks the commission for their hard work. An analysis of the report will be available at spacewhatnow.com.

My Opinion

The day of President Bush's speech, I was asked by a reporter for my opinion of it. I don't think I put my feelings into words well then, and I'm still struggling to do so now. I was in college for the first President Bush's space speech, and turned handsprings over the idea of a mission to the moon by 2010 and a mission to Mars by 2020. Of course, at that time I was going to be a senior in college in the aerospace engineering program at Penn State.

Since then, I've seen the first "We're going to Mars!" charge not even get out of the gate. I've worked in the space industry for close to 15 years, seeing how things work on a day-to-day level. Once, I even experienced the exhilaration of granting the final "Go" for a satellite launch, then felt the ground shake beneath me as a rocket roared to life 1200 feet away, starting its payload's successful journey into orbit. I've watched as the space station has slowly, haltingly, come together through budget and technical difficulties. My footsteps have echoed the halls of Congress, as I've spoken to staff members for my representatives and others, trying to convey my beliefs in the space program to them. I've spoken to the public, at times feeling their excitement while at others catching very barbed questions from sources I didn't expect. I've also heard from another side of the space community, those that think it's time for the government to turn over the reins to commercial efforts.

Facing this "Renewed Spirit of Discovery" with a more experienced eye, I am not turning handsprings. To be sure, I am much happier with the future of America's presence in space than I would be if the announcement hadn't been made. The ship seems to be pointed in the right direction, but there's just the matter of whether or not there's enough wind to fill the sails and push it forward. The wind has to come in three different sources—political/public support, cost, and technical feasibility.

Political and public support should be one and the same in a perfect representative democracy, but today's situation in the United States is far from perfect. In today's situation, it's possible that the space program would be much better off limping along like it was before, mostly under the radar of public opinion pollsters and late-night talk show hosts. Congresspeople with NASA centers in their districts would make sure that enough projects are funded to keep some form of progress going, and life would go on. Occasional successes would bring flashes of news glory, while failures would bring an occasional investigation and usually, in the end, more money to a particular program. Here's an example backing the last statement, in case you have any doubts about it. The failed Mars missions of 1999, attempting to be done for a fraction of the cost of the successful 1997 missions, led to the relatively whopper-project Mars Exploration Rovers of 2003. In interviews, NASA officials actually bragged about the fact that much more money was spent on this project, with the unstated assumption that throwing more money at a mission will make it more likely to work. Conversely, it was the lack of an accident during the 1988-2003 timeframe that led to budget cuts by Congress and cost-cutting measures by NASA that the Columbia Accident Investigation Board cited as one of the causes of the disaster in February 2003.

A Mars mission changes the entire dynamic. Since President Bush's announcement of plans, all manner of debate, both public and private, has taken place and is likely to continue. Newspaper columnists who've never expressed an opinion on space have done so, and in many cases, I believe their opinions are colored by the person (or maybe the political party they belong to) who delivered the policy. I need to step lightly here, because some who read this will think that I'm against public debate, which I'm not. I am against a biased public debate, especially one that masquerades as an unbiased one in the form of a news article by including an occasional "counter-fact" to the article's main point, or fails to acknowledge a reporter's basic slant. An opening line like "I started out in favor of the space program, but as I dug deeper, I found that it was just a pork-barrel project" would go a long way, for me, to give an article the appearance of completeness. I have worked to state my goals up front, and tried to keep my arguments unbiased, though I'm sure that some bias is there.

In the debate, I've heard plenty of the old arguments come up. It's too expensive, there are more important things to do on Earth, why do we need to know about what's going on out there anyway, etc. The expense argument, in my opinion, is largely based on dollar figures that are presented with no sense of context. When people hear about an $820M rover mission, without knowing (or being told) that the money was spent over the course of 4 years, making for an average cost of $205M each year, which works out to an average of 1.4% of NASA's annual budget, which is less than 1% of the total US budget each year. It is tough to focus on such numbers and their size relative to the federal budget when it's possible that, at the same time, their local school board is trying to raise funds to add a wing to their high school and the costs is on the order of $2M. Human nature poses the question of "Why can't that $820M be spent to put a new wing on our school (and 409 others) instead of some mission to Mars?" This is a valid question, and my only answer to it is that, if the space program were completely cancelled, there would still be concerns like this. The argument can be made that there would be fewer such arguments, but I doubt that the $16B, requested for NASA in 2005, would satisfy every need that people in local communities, or members of national interest groups have for this nation.

The need question leads right into the second argument that echoes, in that there are more important things to do here on Earth before we spend lots of money on a space program. Here's my first answer: "I'm glad that Europe didn't set out to solve every problem that they had before they started exploring over the horizon." In the 15th and 16th centuries, as Europe made its first tentative steps towards exploration and eventual colonization of the Americas, the mainland can hardly be described as being in good shape. Plagues, monarchs who weren't quite all together upstairs, and an economic class system that kept the majority of the people poor contributed to a society that seems to me to be a little rougher to live in than the one we live in today. Granted, the problems of the time were not chronicled daily in a news media hungry for ratings, but the problems existed and, for right or wrong, some portion of the day's treasure were spent to find a new route to the Indies. This set off a chain of events that allows me to sit here and type these words on a computer. I for one am glad they happened.

Those who make the argument, however, appear to think that if the space program were cancelled, all of society's problems (or at least a significant portion of them) could be solved with the additional funding. I've expressed my opinion on that theory.

Along those lines, the space efforts of the past were not made in some sort of event vacuum. While President Kennedy's "Special Message to Congress on Urgent National Needs[90]" is most widely quoted as his moon speech, the address called for tax increases for a host of government programs, ranging from the space program to civil defense, Peace Corps projects, and a reorganization of the military.

Technically, I have fewer concerns about this new initiative, because I believe in many ways the fiscal constraints upon it will force some innovation in our ways of doing space travel. One reason that the Apollo program was so expensive was the need to build an entire infrastructure for it—facilities, launch pads, launch vehicles, and tracking stations around the world that we have in one form or another today. This existing infrastructure constrains our actions in some ways, but I don't believe that they need to be overly constraining.

The timeline has me a little concerned. As Norm Augustine pointed out in his testimony to the Moon to Mars Commission, a spaceflight program with a goal of Mars by 2030 has to survive 8 presidential terms, 17 congressional terms, the winds of change in public opinion and potential drastic changes on the world stage. Apollo was originally envisioned as landing on the moon in 1967, when Kennedy could still be in office. Some have pointed to this paradox, saying that if we went to the moon in 1969 after 8 years of development, starting today we should be able to make it to the moon by 2010. If unlimited funds were available, I would concur with that statement, but the political realities of the ISS requiring completion as well as the current budget situation make the idea a non-starter.

As parting words along these lines, I have to paraphrase Robert Zubrin with one of his most challenging thoughts. Who were the King and Queen of Spain when Christopher Columbus sailed in 1492? Can you name any other policy that these monarchs had towards anything? Can you name the monarchs who preceded or followed them? Odds are, unless you're a Spanish historian, you could answer the first question (Ferdinand and Isabella) but not the last two. We are at a critical time, where humankind can choose between two futures. In one future, we move beyond the world that gave us life and nurtured our minds, taking our baggage of social troubles with us, to be sure, but we move. In the other, we sit in place on our world, turn our back on potential lands beyond and resign ourselves to the fact that there are no other places to go. At the same time, we are putting our head in the sand, awaiting an asteroid strike that is preventable. The methods for making such journeys are important, but the decision to go itself is critical. I say we must move on.

For now, I'll watch deliberations in Congress about the presidential budget request for next year. All the while, though, whenever I look up at the moon or Mars, I'll smirk a little bit, thinking that I may yet see people return to one and visit the other in my lifetime. Hoping, all the while, that I may play some part.

Chapter Eleven

The Technical Stuff

It was my goal to remain completely non-technical in this book, so that anyone who wanted to learn about space travel and its promise could pick it up and not be drowned by jargon. Hopefully, I've met that goal. In case someone wants to know a little more about some of the topics brought up in previous chapters, this chapter covers some of the technical explanations that were glossed over for the sake of brevity earlier.

How a Rocket Works

The story is a little old, but a balloon that you blow up and release to fly around the room is essentially a rocket. Granted, it's unguided and of a very low thrust and efficiency (see the next section), but it's more than enough to amuse a room full of kids, or even adults for that matter. The major parts are there, however. A chamber (the balloon) holds a working fluid or gas (the air) under pressure. The nozzle (the hole in the back) allows that gas to escape in one direction, which pushes the rocket in the other direction.

Of course, most rockets that people see are powered by much more than air pressure. Some examples include the Solid Rocket Boosters that lift most of the weight of the space shuttle on its way to orbit, and the Space Shuttle Main Engines that do the rest of the job. These are examples of two types of rockets, solid- and liquid-propelled.

Rockets that run using solid propellants are relatively simple, and they generate a lot of thrust (though not a lot of efficiency, see below). A solid rocket is made up of a pressure vessel with a nozzle at one end. The vessel is filled with propellants, usually made up of a mixture of aluminum and rubber.

When this mixture is burned, pressure and temperature rise quickly within the vessel, and hot gas escapes out the nozzle. Some solid rockets can steer themselves by pointing the nozzle in different directions, though most rely on another form of steering. The ability to steer a solid rocket makes that rocket much more complex.

Liquid-propelled rockets are more complex than solid rockets, but they make up for their complexity by usually being more efficient and by being throttleable (think of the accelerator pedal on your car) and able to be turned on and off again and again. In a liquid rocket, the fuel and oxidizer (called propellants when we talk about them together) are piped to the combustion chamber. There, they are mixed and either explode on their own or are ignited by a spark source. Liquid propellant rockets can be steered rather easily.

Many people think that it's the pressure exerted by expanding gasses that make the rocket engine produce thrust. While this produces a small part of the rocket's thrust, most comes from pushing the mass of burned fuel and oxidizer out the back. Think of standing on a skateboard and throwing a bowling ball in one direction. Assuming that you don't fall over (why I don't recommend you try this at home) you will travel in the opposite direction that you threw the ball.

THRUST VS. EFFICIENCY

This gets a little tricky, but we'll get through it. It turns out that there are two measures for a rocket engine, and anyone designing one has to keep both in mind when building a spacecraft. In some ways they're related, but those ways are pretty strange. I've simplified the discussion here a little, but there are more complete references listed in the back if you want more information.

First comes thrust, which is what most people think of when they think of rockets. Saturn V, the rocket used to take people to the moon, had 7.5 million pounds of thrust. Of course, the rocket weighed 6 million pounds at liftoff. Thrust is essentially a measure of the rate (in units of mass/second) that the rocket in question can throw its burned propellants overboard. This is one reason that solid rockets, with all their problems listed above, are worthwhile in vehicles launched from the ground. Solid rockets can get such a high mass flow rate that their thrust is very high for their size.

Next comes efficiency. The most important measure of efficiency in rocketry is specific impulse, abbreviated Isp. Specific impulse is defined as the thrust that can be obtained with a propellant mass flow of one unit per second. Each propellant type has its own value for specific impulse. Typical numbers range from 300 for kerosene and liquid oxygen, 450 for hydrogen and liquid oxygen, 900 for a nuclear rocket, and 10,000 for ion engines. Bigger numbers are better. Because of the rocket equation (I fought hard to keep this book "equation free" and resisted the temptation to put the rocket equation in), a spacecraft that has half its mass in propellants can accelerate to a speed of a little less than 7 times the propellant's Isp value. So, a kerosene rocket made up of half fuel could accelerate to 2100 meters/sec, while the ion engine could accelerate to 70,000! So we should just put ion engines on all spacecraft, right? Not so fast. While extremely efficient, ion engines are extremely low in the thrust department.

So now that we know the two measures of rocketry, let's see how they apply in the environments that rockets operate in. Here there are essentially two types: against a gravity field (and usually in an atmosphere) or perpendicular to a gravity field. When going against a gravity field, like right after leaving the ground, every second that the rocket is firing the entire craft is being pulled straight back down by gravity. There's also the effect of the atmosphere trying to slow it down, but gravity is the big concern here. In this situation, a rocket wants to fire very quickly, and must always exert more force than its mass times the gravity field it's in. This type of application favors big rockets with large mass flow rates. Specific Impulse, while something to keep in mind, is not the main concern. The Saturn V used kerosene and oxygen as its first stage propellants, not because of its specific impulse, but because of its mass flow rate. There were other factors, like the fact that the largest engine at the time used kerosene and oxygen, but that's beyond the scope of this section. Imagine using an ion engine in a gravity field. As described, the ion engine is extremely efficient, but has very little thrust. In fact, its trust is so small that an ion engine cannot even lift itself in an Earth gravity field. Once you get above the planet's atmosphere, and are traveling perpendicular to the gravity field (like a spacecraft in orbit), then the situation changes.

Orbiting spacecraft are not directly fighting gravity, and have no (or very little) concern with atmospheric drag. Therefore, their propulsion options are much better. In this case, thrust is not as much of a factor, and efficiency becomes very important. Ion engines come into their own in the vacuum of space, thrusting for weeks or months to very slowly accelerate their craft at rates where the distance portion is measured in millimeters. By the end of thrusting, the craft is going much faster than it was, and used very little propellant.

Remember, I've simplified a lot of facts to cover their explanation in these small sections. Ion-propelled spacecraft require quite a different trajectory than chemically propelled craft, and entire careers have been spent working on the subtle approaches that can squeeze a little more velocity out of the same amount of fuel. Also, I've given no treatment to the extremely high temperatures and temperatures that exist inside traditional rocket engines. These devices are by no means simple...but I believe that they don't need to be as complicated as they're currently made and used.

Orbits

The goal of most rockets launched off the surface of the Earth is to make some object orbit our home planet. Even when a spacecraft is launched beyond Earth orbit, such as a probe to another planet or to fly around the sun, that probe orbits the Earth for some period of time. So, we talk about orbits and orbiting a lot, but what's involved?

The concept of putting an object in orbit around a planet was actually discussed by Sir Isaac Newton in the 1600s. He considered the idea of placing a cannon on a high mountain, and firing the cannonball parallel to the surface. At low speeds, the cannonball would fall to the surface, but, he surmised, if the ball could fly fast enough, it would fall at a rate equal to the curvature of the Earth. At this point, the ball would be in orbit. I don't believe he covered the topic of whether or not you'd need to duck 90 minutes later as the cannonball came back around, but that's not a concern, and we'll get to that in a moment.

Getting into orbit is a big deal. That's the reason we build (and will continue to build for the foreseeable future) those huge rockets that everyone loves to watch. Rockets burn a whole lot of fuel and expend all that energy to get about 5% or less of their initial mass into orbit. To achieve orbit, an object has to be above the Earth's atmosphere, and be moving at about 7 km/sec (that's about 16,000 mph!). Because of the gravitational pull and drag effects described earlier, a rocket must actually do the work to accelerate to 9 km/sec on its flight to reach altitude and achieve the 7 km/sec rate. For more detail on rocket engines, see the previous section. If you want to know more about getting to orbit, hang on and we'll talk about it in "Getting to Orbit."

Once you get into orbit around the Earth, though, staying there is pretty easy. While Earth's atmosphere slows you down over time, and will bring you back home if you don't do anything, as long as you're about 100 miles or more

267

above the surface, you can rely on at least a couple days in orbit. The only time you really have to expend rocket fuel in orbit is if you want to stay up for a while, and counteract some of the work the atmosphere did to pull you in, or if you want to change your orbit. Here's another spot where we're going to get dangerously close to some math.

Let's start by saying that we're orbiting the Earth in a circle, and we're supposed to go to the new supply depot, recently launched by some brilliant entrepreneurs. Our rocket just dropped us off, but we're a little low compared to the depot, which is in a circular orbit some kilometers above us. That's OK, we brought some fuel and small rocket motors with us so we're in good shape.

If we want to go higher in orbit, we need to push ourselves (using our craft's rockets, of course) in the direction we're already traveling by firing our rockets opposite to the direction we're moving. This speeds us up, and will take us higher above the Earth when we reach the opposite side of the planet. That gets us closer to the supply depot, but we're not there yet. It turns out that firing a rocket once while we're in orbit puts us into an egg-shaped orbit (again, the technical term is ellipse, and in that same vein, there's no such thing as a perfectly circular orbit, either, but you can get close!) where we alternate between being far and near the Earth. Remember, we said that the depot was in a circular orbit above us? So, we've got to fire our engines one more time when we reach the other side of the world. (Don't worry, it only takes about 45 minutes when you're in low Earth orbit.) The second rocket burn, pushing us in the same direction we're traveling, puts us in a circular orbit at the same altitude as our fuel depot. Now, properly timed, this little two-burn maneuver could take us right to the depot, but space folks like to be a little more conservative. What they'll normally do is put an approaching craft a set distance from the depot, drifting towards it very slowly. Then, when they get REALLY close (like in sight of it) they'll do another maneuver or two called trim maneuvers to put the two craft in exactly the same orbit.

When it's time for us to come back to Earth, or just lower our orbit, we simply fire a rocket engine pointed in the direction we're traveling. This slows us down, and lowers our altitude at the far side of the Earth. If we slow down enough, we'll touch the upper levels of the atmosphere, and if done properly, that atmosphere will slow us down even further for a return to the surface.

Changing our altitude in orbit is pretty easy and doesn't cost us much propellant. There's another way to change orbits, though, that takes much more fuel. In fact, it takes so much more fuel that great measures are taken to keep from having to change orbits in this way. This change is called a change in inclination. Inclination will be discussed later (see "Getting to Orbit"), but we need to talk a little bit about it here.

Every orbit has an inclination, that is, an angle it makes with the Earth's equator. Some example inclinations include geosynchronous satellites staying very near zero, the space shuttle on a mission to repair the Hubble Space Telescope at around 28 degrees, and the International Space Station is located around 52 degrees.

Now, you'll remember earlier when I said that objects in orbit have to travel about 7 km/sec to stay there. Well, it turns out that if you want to change your inclination, you have to point your rockets roughly sideways compared to the direction you're traveling. As you can guess, this doesn't really speed you up or slow you down, but simply changes your direction. For instance, if you want to change your inclination by 60 degrees, you'll need to fire your rockets enough to move 7 km/sec! As a reminder, that's the amount of speed you needed to achieve with your rocket earlier. Now, a sixty-degree inclination change is a little extreme (though some Russian communications satellites have to go through 52 degrees, which is one reason their rockets are so big!), but let's take a look at a more practical case, and one that's drawn a lot of attention since February 2003…a space shuttle going from a non-space station orbit to a space station orbit.

The space shuttle *Columbia* was launched into an orbit inclined approximately 39 degrees to the equator. The ISS was at its usual duty station, inclined 52 degrees to the equator. Now it's true that news stations (and NASA) made a big deal about the fact that the two craft passed close to each other a couple times, and could even see each other out their windows, but this is a little misleading. If you're driving on a freeway that leads from north to south, and someone else is driving on a freeway that leads east to west, you may actually see each other for a moment as you pass over/under each other. This close approach doesn't mean that your journeys were anywhere near each other.

So, the space shuttle and the ISS were about 13 degrees apart in inclination. They were both traveling at about the same speed, and for one to move to the other's orbit, they'd have had to change their inclination by those 13 degrees. Again, I'll save you the math, but it works out to be over 1.5 km/sec. The space shuttle cannot carry that much fuel on board. It's true that shuttles can carry extra fuel if they know that they need to maneuver a lot as part of a mission, but extra fuel that they bring along means less cargo they can take up.

Tom Hill

Reusability

If there is a holy grail of space technology right now, it would have to be reusability. Reusability, properly implemented, will bring down the cost of flying to low Earth orbit (barring any radical new capabilities or planetary-scale projects like space elevators or rail guns), and it has the potential of lowering the cost enough to really make spaceflight something that can happen every day. "Hold it!" you're thinking. "The space shuttle is reusable, and this guy spent a lot of time describing how it didn't bring down the cost of launch." All that is true, and I won't back away from it. I will point out, however, that the space shuttle was designed with a lot of cost savings in mind in initial production, in many ways causing high operational costs down the line. This type of mindset must change in a reusable vehicle in the future.

In many ways, I do not disparage NASA for building the shuttle in the way they did. Based on the budget climate of the time and the promises that were made in order to get what funding came for the spacecraft, designers and builders had little option other than cutting some things extremely fine. The unforgiving nature of space travel also played a major role.

So the space shuttle has some problems with how it implemented reusability, but when you translate that term into its aircraft equivalent, maintenance, so did the Boeing 707 and all early jet aircraft. The important thing is that we learn from the shuttle, and build a better reusable spacecraft next time around. Unfortunately, spacecraft are built so rarely in today's paradigm that we'll need to incorporate a lot of lessons in the next version, as opposed to Boeing having time to learn lessons over multiple aircraft such as the 727, 737, and so on before their masterpiece of today, the 777.

Why is reusability still the goal? Because properly implemented reusability will lower the per-launch cost of a mission to the propellants and a portion of the personnel required to maintain the flight rate. The cost of propellants is pretty much non-negotiable except for buying in bulk, but the personnel cost will go down dramatically as more missions per year fly. If 100 people work on a crew, making an average of $50,000, labor costs (dangerously, we're ignoring overhead here, likely up to 3X the salary cost of workers) are $5M a year. That stays the same whether this operation launches 1 mission a year or 100. If only one mission flies, then labor costs for that mission are $5M. Fifty flights a year bring the labor cost down to $100K per flight.

How is reusability achieved? Well, there are a couple ways, but both of them involve building hardware on a scale like we've not seen before in space travel. One route that can be chosen is to design and build the hardware a little simpler and a little more robustly. The space shuttle main engines, for example, have to have every inch visually inspected between every flight. On every fifth flight, the engines have to be taken apart for a major overhaul. This is not the case for aircraft engines. What probably happened with the shuttle engines is that a certain performance level was promised, and then as development proceeded, engineers realized how difficult their promises would be to keep. From here, design compromises were made that cut into the "beefiness" of the engines themselves. Now, engines without "beefiness" can be safe, but they have to be watched much more closely than a beefy engine. By building future engines more along the lines of aircraft engines rather than the space shuttle's, such long, drawn-out inspections and frequent overhauls will not be necessary. Based on this operation style, a reusable craft returns from its flight and receives a simple spot check of some critical components or relies on the fact that there were no problems on the flight to certify it for its next mission.

It's possible that making tougher components will only take us so far in reusability. Tougher engines are heavier, and there are some laws in physics that we haven't found a way around yet, which say that having heavier engines means that we need to carry more fuel, less payload, or both. Therefore, another option is to just build a lot of engines that require more maintenance than aircraft engines (but less than shuttle) as a goal, and then have these engines in a rotation that keeps a number of them ready to go at any one time. Using this model, a reusable spacecraft returns from orbit, automatically has its engines removed and new ones installed. The "old" engines go to the engine shop for some form of maintenance and then get put in line to be used again when their turn comes.

The second method discussed is obviously less efficient as far as costs go, but if the flight rate of this mythical vehicle were high enough, the costs should spread out and not be a major factor. It's likely that some sort of compromise between the two will have to be struck, where engines are OK to fly 5 times without removal, but then have to be taken out for inspection and maintenance. At today's flight rates, it would be hard to justify a standing shop for working on engines given that rotation, but if flights were happening many times a day, the need becomes clear.

Getting to Orbit
(Why can't I fly from Florida and go over the poles?)

It isn't widely known, but the United States has two major launch sites. No, I'm not talking about the fact that NASA and the US Air Force share the thin strips of land on the East Coast of Florida. The launch site that gets all the press is Kennedy Space Center/Cape Canaveral Air Force Station in Florida, but there's another range that holds launches as well. It's called Vandenberg Air Force Base in California. Why do we need two launch bases? Well, it has to do with the direction that we're willing to launch our rockets. By the by, I'll probably catch heat from Wallops, VA, Kodiac Island, AK, Black Rock, NV, and any number of other locations where people have sent rockets into the atmosphere and beyond. For now, we'll confine our discussion to locations in the US that regularly send materiel into orbit.

The United States will not launch a rocket in a direction that takes it over human populations during its first minutes of flight. I say the first minutes because eventually, anything launched into orbit will pass over people, but by then the threat it poses is minimal until the time comes to return to Earth. No, the primary concern here is when the heavy hardware is belching fire and accelerating away from the launch pad. This decision limits the US in the directions it can launch in to flying over water. Other nations, such as the former Soviet Union and China, decided that risk to human life was minimal from these rockets, and chose to put their rocket centers within their borders. For the most part, this decision seems to have caused relatively low loss of life, but a Chinese launch in 1996 veered off course and crashed, and may have destroyed an entire village. The Chinese have not been very forthcoming with a damage assessment from the incident.

So starting with the idea that flights can only take place over water, launches from Florida are limited in their direction by the rest of the East Coast to the north, and by the Caribbean islands to the south. Vandenberg has its limits as well, and can launch on either side of due south, or west. These directions of flight, called vectors or azimuths, limit the orbit inclination that a spacecraft can get into from each launch site. Remember from the previous discussion on orbits that once an inclination is chosen, a spacecraft would have to spend a lot of fuel in order to change it.

With some fancy flying, and the expense of a lot of fuel, a mission can fly into a different inclination than is normally reachable from a launch site. These "dog-leg" maneuvers have the craft liftoff and fly along a normally acceptable trajectory, then turn (usually left when viewed from above, but of course directions like that don't mean much in space travel). Maneuvers like this provide quite a show for the US East Coast at times, as the space shuttle or other launch vehicle flies along the coast.

Planetary Launch Windows
(Why can't I fly to Mars whenever I want?)

We just finished a discussion as to where a spaceflight can go to based on where it starts on Earth's surface. There is another variable at work in this discussion, and that is time. Time comes into play because once a rocket launches from the surface and enters orbit, in most cases, that orbit stays in the same position that it was for the spacecraft's life. The Earth rotates below the spacecraft, showing a different portion of land each time the satellite goes around the planet. Therefore, the time that a satellite launches can be very important in what it will see during its lifetime. Essentially, the rocket sits on the pad until the Earth rotates under the planned orbit for the satellite. When they line up (or come very close), the rocket takes off and flies into the desired orbit, setting the location for the orbit. This is why some launches take place at seemingly odd hours of the day or night. The time period where a rocket can launch into the proper orbit is called a launch window, and if you think they're tough for launches into Earth orbit, wait until you hear what's involved with traveling to another planet!

As humans realized for a while, then forgot, then rediscovered in our past, the Earth travels around the sun. I hope I'm not stretching the point too far to say that other planets orbit the sun as well. Because the planets lie at different distances from our star, they travel around the sun at different rates. This leads to a racetrack-like situation where the planets closer to the sun whip around relatively quickly (Mercury, the first on the way out from the sun, takes only 88 days to make its transit) while planets further out take a much longer period of time (Jupiter takes nearly 12 years and Neptune takes about 164). Based on our technology today and for the foreseeable future, spacecraft that are going to fly from the Earth to another planet have to take these rotation times into consideration when they decide to fly a mission. Earlier, we discussed a spacecraft in Earth orbit changing that orbit to rendezvous with an orbiting supply depot. Well, the same approach must be taken to travel from Earth to another planet, only the values of velocity

273

changes are much higher. Essentially, for a mission to fly from Earth to another planet, we must wait for that planet to be in such a spot that it will be halfway around the sun from where Earth is when the mission is launched. This time is called a launch window, and can be open for very short periods of time (on the order of seconds) to long periods of time such as months. When our craft arrives at its destination, it must then fire rockets or use the planet's atmosphere to adjust its speed in order to stay.

SLINGSHOTS

In space missions, you'll often hear about a spacecraft making a close approach to a planet to "slingshot" its way further out into space. This is a trick that's been used since the '70s and some missions have made use of several slingshots to make their mission possible. How does this work? As usual, we'll start with an example.

Take a rubber ball and throw it straight down on the ground. If you have any skill at all, the ball will bounce right back up to near your hand, and you'll be able to catch it again. You can do the same thing against a wall, although this throw may take a little more skill. Now, while it's true that there are some losses when the ball bounces, it leaves the wall with almost exactly the same energy that it arrived with.

Let's shift the vision, so that now you're bouncing the ball off the front of a box car in a railroad yard. If the car is standing still, the ball bounces away from the car the same as it would bounce away from a wall. If the car is coming towards you, and you throw the ball against it, when the ball returns to you (just before you jump out of the way of this moving car, of course) it is moving faster than when you threw it. The reason is that while the ball was in contact with the moving car, it picked up some of the car's momentum. The ball moves faster after the collision, and the train car moves a little slower. If you were to throw the ball against a train car moving away from you, the ball would actually give the car a little push, and would bounce back towards you at a lower speed than when it hit the car.

Let's translate that into space travel. In orbits earlier, we discussed the fact that if you are traveling in an elliptical orbit around a planet, you'll move faster as you approach the surface and slower as you move away. This is similar to the original example of bouncing a ball off a non-moving wall in that everything stays the same for the most part. If your orbit takes you close enough to the planet to make you graze the planet's atmosphere, you'll lose some energy and your ellipse will be smaller, but that's getting a little complicated for this section.

If, on the other hand, you start out orbiting the sun, things get a little more interesting. As your spacecraft approaches the planet, the planet is moving relative to your orbit. Therefore, the planet serves as the moving railroad car, and can change your velocity (either your speed or direction, but usually a mix of the two). Here the interaction is a little more complicated, because our ball bounced off the railroad car for just a fraction of a second, but the planet will significantly affect our sample spacecraft for days to months depending on how large the planet is. This complex interaction gives us some great options that can make our missions quite flexible. In the case of the Cassini mission to Saturn and the Galileo mission to Jupiter, it allowed a much larger spacecraft to be launched from Earth than would have been possible without gravitational slingshots. A mission named *Ulysses* swung around Jupiter to move its orbit relative to the sun's equator, giving us unprecedented views of our star's poles. In the mother of all slingshot stories, the *Voyager 2* spacecraft used 3 precisely guided slingshot maneuvers to give us a "grand tour" of the outer solar system. This little emissary robot visited Jupiter, Saturn, Uranus, and Neptune.

There is a current move within NASA to stop using the slingshot effect, because it sets dates that you can and can't launch due to planetary alignments with the planned slingshot planet. This is part of a larger effort to "free us" from launch windows altogether. The reasons cited in the argument relate to the fact that slingshots and launch windows are old technology, and we need to move beyond that old technology in order to move into the solar system. In my opinion, this goal, while sounding good on the surface, is misguided. The fact is that launch windows will always be with us. Even when launching to Earth orbit, most launches are very particular as to when they can launch. The reasons range from spacecraft design (some spacecraft cannot launch at certain times of the day or year because it would put them in Earth's shadow for too long a period) through mission plan (some space missions want to fly over a location on Earth at a certain time of day each day, and to do so they must get launch right around that time) but they will always be there, and trying to say that we don't want launch windows in solar system exploration is very similar to saying we don't want launch windows in Earth orbit missions. I'm not saying that there aren't ways to minimize our need for launch windows. There are propulsion systems such as nuclear-powered ion engines (see the next section) and solar sails that don't rely on launch windows. In many cases, though, because of the complex dance that describes planetary motion, these propulsion systems will actually lead to a longer period of time traveling to the objective planet than the "low tech" launch window or slingshot approach.

I believe that gravitational slingshots, in one form or another, will be with us for some time. In fact, Robert Zubrin describes a modified slingshot[91] around the planet Jupiter that would allow us to do some interstellar exploration with relatively low-tech propulsion.

NUCLEAR POWER IN SPACE

In earlier discussions about nuclear power in space, I stated the fact that they were necessary and moved on. Now I'll give the reasons why.

First, an overview of nuclear power in space is required. Here, I describe nuclear power as any device that relies on the decay or splitting of heavy atoms for energy. Right now, all activities in space that use nuclear power rely on the decay of atoms, but in the future, it's likely that the splitting of atoms will take center stage.

Many large atoms are unstable. Over a period of time they will split into smaller, more stable atoms. This action is called decay, and it takes place at a rate specific to each atom in question. This rate is described as a half-life, in that after one half-life passes, one half of the material has decayed into lighter atoms. When the heavy atom splits, energy is released in the form of heat and some form of radiation. The heat is used either in its direct form to warm things, or is converted to electricity to power a craft. If only heat is desired, small patches called reactor heater units or RHUs can be placed in different spots on the spacecraft. Using these RHUs and insulation, a spacecraft can use a minimum of power and still keep itself warm, even through a long cold Martian night.

For electrical power, radioisotope thermoelectric generators (RTGs) are the nuclear system du jour. These units use the heat provided by an isotope's decay to produce electricity through a device called a thermocouple. Because this system has no moving parts, it is extremely reliable. The *Pioneer* 10 spacecraft was powered by RTGs and ceased operation just recently (actually, what happened is that Earth tried to contact it recently and didn't get an answer. It's likely that the probe is still functioning somewhat, but unable to broadcast with enough energy for us to hear it), having run for more than 30 years.

Eventually, even RTGs will reach their limits. Despite being very reliable, they don't produce a lot of power—Cassini, the Saturn mission, runs on 700 watts, about the power needed to run a microwave oven. As humans require power to keep them alive beyond low Earth orbit, thousands of watts will be necessary, and for that we'll need to actively split atoms in a process known as fission. Fission reactors are in common use today on the planet Earth. In the reaction, a heavy atom such as uranium is split into two lighter atoms, but at a much higher rate than happens in natural decay. This higher rate produces more energy, but also more radiation. The energy will be quite useful, but the radiation will simply have to be dealt with. Nuclear fission reactors, placed a good distance away from a crew (or well shielded, such as on a planet or moon), will provide the energy for exploration. Fission also opens the way for nuclear thermal rockets, or NTRs, that will provide more boost for a unit of propellant (higher Isp, remember?) than our current chemical rockets.

Space missions within the inner solar system are kept alive and powered, for the most part by solar energy. I say for the most part because two Mars landers, *Viking* 1 and *Viking* 2, were powered by RTGs, as were a number of lunar-surface missions, and many other missions carry the RHUs to keep sensitive equipment warm when the environment gets really cold. Solar power, converted to electricity by solar panels, can provide enough energy to run a spacecraft and allow it to do some great things. Of course, more power would allow the craft to do more things, but we'll get to that in a moment.

As distance from the sun increases, the overall power provided by the sun decreases. This decrease is not linear, going down in a straight line, it's exponential, meaning that it goes down by the square of the distance. For example, if you were at some distance from the sun and received 4 units of energy, and then moved to twice the original distance from the sun, you would receive 1 unit of energy. If you move ten times the original distance from the sun, you would receive .04 units of energy, and so on.

How does this translate to solar system exploration? Mars is 1.5 times as far from the sun at Earth, and it receives, on average, about 40% of the energy from sunlight as Earth does. Jupiter is 8 times farther, and receives about 2%, and Saturn is 15 times farther, and receives less than one percent of the solar energy Earth does.

Another factor is the period of time that a mission might go without seeing the sun. Our Earth orbit satellites can sometimes lose solar power for 75 minutes at a time. This provides quite a design challenge to spacecraft builders, because in many cases, the satellite must continue all its functions even though it lost its primary power source. Earth-serving craft make up for this dark time with batteries that drain while the satellite is in shadow and then recharge them while the sun powers their systems. The batteries are recharged once the sun returns to the satellite. A craft on the surface of Mars will suffer 12 hours and 20 minutes of darkness, while one on the moon will meet up with 14 Earth days of dark time. Current missions like the Mars Exploration Rovers get around this problem by shutting down at night, but future missions, especially those involving humans, will want to keep functioning through the night. The battery packs required to keep a craft powered for these intervals are so large as to be impractical, and the huge solar arrays necessary to recharge this system after a period of darkness would be at least as impractical. In contrast, nuclear power sources provide power all the time, not subject to outages due to lack of sun.

Nuclear power also allows radically different missions than the ones we're used to today. The huge increase in electrical power allows a proportional increase in data return, because radio transmission datarate is directly dependent on the power of the transmitter. These new missions will have a lot of data to send back, too, because their increased power supply will allow them to actively scan a planet. As amazing as the data is that we get back from our spacecraft today, most of it is based on energy coming from the planet being scanned. For example: visible-light images (plain old pictures) rely on light from the sun reflected off the planet, maps of water-bearing ice from the moon or Mars rely on cosmic rays, and magnetometer readings rely on a planet's existing magnetic field (or lack thereof). Some of the best information we've received comes from spacecraft that actively send out signals and then read them back. Here an example is the map of Venus produced by the Magellan spacecraft, and the graphic relief map of Mars produced by the Mars-Orbiting Laser Altimeter on board the *Mars Global Surveyor*. With nuclear power, the options for such active scans grow by leaps and bounds allowing, for example, active radar scanning of a planet surface with different frequencies (giving different returns based on the minerals on the ground).

As it turns out, nuclear power sources not only provide electricity, but also provide heat. Heat is a very good thing for people (we tend to get a little cranky below 20 degrees Celsius or 65 degrees Fahrenheit, some people get cranky below 70). The waste heat provided by RTGs is enough to keep its spacecraft from freezing in the ever-colder reaches of the outer solar system, while the waste heat from an actual nuclear reactor would be enough to keep a colony warm.

Currently, there is an effort underway at NASA to send a nuclear reactor into space. Project Prometheus, begun in 2002, will develop nuclear technology for space exploration. It's likely that this project will receive a boost as part of the new space initiative, because of the critical role that reactors will play in human spaceflight. The first mission, as currently planned, will attach a nuclear reactor to a group of ion engines and send the spacecraft to Jupiter, and be called the Jupter Icy Moons Orbiter (JIMO). Theoretically, this nuclear-powered craft would be able to leave earth at any time, fly to Jupiter at its leisure, and then use its near-infinite supply of power to orbit many of the interesting moons found there. In its current implementation, however, this mission is unlikely to succeed. Currently, in a desire to avoid launch windows, the mission is designed to thrust its ion engines the whole way to Jupiter. With the huge reactor planned for flight, this mission will take only 7 years as opposed to the 12 years it would take a small reactor to get there, but that's 7 years that the nuclear reactor must run at full power and the ion engines must operate without a hitch. This is a long time for both the first nuclear reactor to fly in space and for a bank of ion engines. Yes, missions have run for seven years in the past, but most of the time, long periods are done with the craft in some sort of "quiet" mode where all of their systems are not up to full power. Here, the real mission starts after spending seven years at full power! Ironically, the quiet times come when using the "old-fashioned" and "low-tech" methods of traveling like launch windows and slingshots.

Unfortunately, with the possible shaky ground that a nuclear initiative is on due to political fallout, any sort of failure for this mission may prove difficult to overcome in the future. Let's say that the reactor goes like a champ for six and a half years, and then something crops up that keeps the mission from making it to Jupiter. Given the likely court challenges to take place before launch, what scientist will want to put his or her experiment on top of a nuclear reactor again in the future?

The alternative is rather simple, but it appears as though bureaucratic momentum (one of the most powerful forces in the universe, I've found) is moving towards the more complex solution. If we accept the fact that launch windows will be with us, in one way or another, forever (I hope I've made that case) we can launch a JIMO mission that's much smaller, lighter, and simpler than those envisioned today. In this alternative plan, the reactor is launched from Earth cold, just as before. The spacecraft is sent to Jupiter on a traditional ballistic trajectory, arcing its way out to the gas giant planet over the course of 2 years. If the craft gets too large to go there directly, a slingshot around Venus or other planets can be used, as was done for Galileo and Cassini. On its journey, the craft operates using its reactor as an RTG, generating electricity off the waste heat of decay from the elements within. As the craft arrives at Jupiter, the reactor is revved up and put into action, driving whatever power system is necessary for the mission. In this mission profile, the craft arrives at its target planet with a probable seven years of life left in its reactor, and ion engines ready to thrust and take the mission wherever it'd like to go. This mission plan demonstrates new technology where it's necessary without overselling its potential. Recent news indicates that NASA may be re-thinking this approach.

So, in the end, nuclear power in space is critical for future exploration, but I'm concerned with how it's being implemented. If I'm ever on a planet or moon facing a 12-hour or 14 Earth-day-long night, I'd like to have a power source that's always ready to supply my electrical needs. Until nuclear fusion comes on line (and becomes much simpler than envisioned today), nuclear fission is the way to go.

Epilogue

Although my experience base in the writing trade is small, I believe I already share a feeling that all authors have as they approach a publisher's deadline for a current events book: the dread of being overcome by events. In the delay between manuscript submission and publication, the world can change around the book as it moves through editing, layout, and uniquely publishing-related preparations towards that time when the rest of the world gets a chance to admire the author's brilliance, laugh at his arrogance, or do something in between. On-demand and electronic publishing can cut down the time it takes for a book to move from the author's computer to marketable commodity, but the larger processes just mentioned will always slow publication.

When I started writing this book, I didn't have a publisher, and had no idea whether I was spending time on a project that would even get anywhere. As fate would have it, the first publisher I contacted accepted my work, so I poured my efforts into completing the best book that I could by their deadline.

Since that deadline, I can't imagine a more turbulent five months in space travel where so much has happened. Some events were on the horizon, or at least predictable as the main book came together, others came seemingly from left field, and all of them will shape the next years and decades of human achievement and existence in space. Here's a mere sample at what has taken place since March of 2004:

- The Mars rovers have made amazing discoveries about the water history of Mars, and have more than doubled their original life expectancy. They are only now starting to show some signs of age. The Jet Propulsion Laboratory has hinted at the idea that the golf-cart-sized vehicles may survive the winter on Mars. These two vehicles, combined with the documented lifetime of the *Viking* landers, put to rest many concerns that human-built hardware will have a difficult time surviving in the dusty environment of the Red Planet.

- The X-Prize has been won by Burt Rutan's *SpaceShipOne* craft. In the midst of the press hoopla surrounding the flights, Sir Richard Branson of Virgin Industries licensed the technology used for *SpaceShipOne* and asked Scaled Composites to build five 6-person craft that will start flight in 2007. In one of the more exciting portions of his announcement, he stated that all monies made from the effort will be put back into space tourism. Other teams say that they will still fly their craft, and the X-Prize foundation is moving on to new challenges, offering prizes for both technological and social problem-solving.
- Two separate lines of research, one Earth-based and one based on information from a craft in Mars orbit, agree on the fact that there is methane in Mars' atmosphere. Methane only exists in Mars' atmosphere for a short period of time, and measurable levels today indicate the potential of very recent volcanic activity, or the possibility of existing life on Mars. There are also hints that more interesting chemicals, even more closely tied to life (ammonia, formaldehyde) have been detected but the scientists in question want to repeatedly verify the information before officially announcing their results.
- The Vision for Space Exploration, as President Bush's January announcement came to be called, has met mixed reviews. Op-ed pieces across the nation accused the President of everything from proposing a one-trillion-dollar joyride to playing politics, although others point out that the idea of retiring the space shuttle in the year 2010 means a loss of jobs in Florida, a state that President Bush is likely sensitive about. The White House broke with tradition when, after Congress cut funds from NASA's budget set for starting work on the vision, the Executive Branch threatened a veto of the spending bill unless the funding was restored. The reason this step is unprecedented is that NASA spending is grouped with Veteran's Administration and Housing and Urban Development budgets, and veto threats to these budget items are rare.
- In the space activism chapter, I recommended that all space advocacy organizations unite in order to achieve the results that all of them are interested in. This has essentially happened. The Space Exploration Alliance is currently a group of 21 space advocacy groups who share in the idea that the vision for space exploration is necessary for the future. A Moon-Mars Blitz, where individual members of the Alliance met with members of Congress was held on Capitol Hill on July 11th-13th, and I participated in it. A group of industries have formed the Coalition for Space Exploration, and there are some advocacy groups that are members of both organizations.

- NASA held its first workshop for Centennial Challenges, the space agency's prize effort. More than two hundred people attended and submitted a wide range of prize ideas and rules suggestions for prizes. Within Congress, the Centennial Challenges program has received a chilly reception, possibly because it is interpreted as part of the vision for space exploration.
- The *Genesis* return capsule, designed to deliver precious samples of interstellar space to Earth for analysis, did not deploy its parachute and slammed into the Utah desert. While press reports about recovering the samples remain upbeat, a preliminary investigation points to the possibility that poor design or construction may be the cause of the failure.

I see no reason that the next five months will be any less turbulent. The 2004 presidential election within the United States will play a major role in shaping future space policy. Success of the X-Prize competition will impact corporate investment in all space activity, and the *Cassini* spacecraft, orbiting Saturn, will dazzle the public with images of the ringed planet and its moons. To keep up with these events, this book's website, http://www.spacewhatnow.com, will chronicle them through links to news articles, occasional commentary by yours truly, and insight into my latest projects. As Burt Rutan said, the next thirty years are going to be quite a ride, so hang on and let's go!

Tom Hill
October 2004

Glossary

AAS	See Alternate Access to Space.
Alternate Access to Space	A NASA program designed to allow commercial interests an opportunity to deliver low-dollar value cargo (like food, water, and oxygen) to the ISS.
Apollo	A program mainly known as the three-man spacecraft flown by NASA from 1968-1975, though it incorporated the lunar module and later efforts such as Skylab.
Asteroid	A rock or mini-planet that's part of the solar system. The majority of asteroids lie in an orbit between the planets Mars and Jupiter. Over time, some have drifted from that location becoming near Earth asteroids or near Earth objects.
Atlantis	Space shuttle orbiter.
Atlas V	The Evolved Expendable Launch Vehicle rocket built by Lockheed-Martin.
ATV	See Automated Transfer Vehicle.
Automated Transfer Vehicle	Project of the European Space Agency, designed to be launched on top of the Ariane V launch vehicle and carry supplies and small equipment to the ISS.

Boeing	Major military and space contractor. Known for building passenger aircraft and launch vehicles such as the Delta series, now serving as part of the Evolved Expendable Launch Vehicle.
Buran	Space shuttle orbiter built by the Soviet Union. Flew one time for a single orbit. Looks very similar to the US shuttle, but relies on its core vehicle, named Energiya, for thrust. The US shuttle carries its own engines.
Cape Canaveral Air Force Station	A US Air Force station from where most US launches take place. People have not flown out of CCAFS since the Gemini program in the 1960s.
CCAFS	See Cape Canaveral Air Force Station.
Centre National d'Etudes Spatiales	The French space agency. Largest contributor to ESA, and major developer of the Ariane launch vehicle.
Ceres	First main-belt asteroid discovered. Larger than most moons in the solar system.
CEV	See Crew Exploration Vehicle.
Challenger	Second shuttle orbiter to fly. Lost with its crew during launch in January 1986.
Clementine	A Department of Defense mission designed to fly to orbit the moon and then carry on to an asteroid in an effort to test ballistic missile targeting methods. Spacecraft gave indications of ice at the lunar south pole.
CME	See Coronal Mass Ejection.
CNES	See Centre National d'Estudes Spatiales.

Columbia	First space shuttle orbiter to fly. Broke up on reentry and was lost along with her crew of seven in February 2003.
Comet	Described as a dirty iceball, or icy dirtball, comets are thought to be leftovers from very early in the solar system's history. Most exist beyond the orbit of Pluto, in either the Kuiper Belt or the Oort Cloud. Occasionally, an object's orbit changes and becomes a comet, making regular visits to the inner solar system and taking on its trademark tail.
Coronal Mass Ejection	Cloud of gas released from the sun. Contains charged particles that can be dangerous for a crewed spacecraft, but also has impacts on Earth.
Crew Exploration Vehicle	Replacement for the Orbital Space Plane, this vehicle, as proposed by the President, will serve multiple purposes as both a space station ferry and integral part of missions to deep space. Scheduled for crewed flight by 2014.
Dactyl	Moon of the asteroid Ida. Discovered by the *Galileo* space probe.
Delta IV	The Evolved Expendable Launch Vehicle rocket built by Boeing.
Deoxyribonucleic acid	Building blocks of life on Earth.
Diemos	Moon of Mars.
Discovery	Space shuttle orbiter.
DNA	See Deoxyribonucleic acid.
Dry mass	The mass of a spacecraft after it has exhausted its fuel. The ratio of dry mass to the craft's wet mass is instrumental in how fast a spacecraft can go.

Earth Return Vehicle	Part of Mars Direct. The ERV flies to Mars before the crew, then generates propellants for the crew's return and other activities on the surface.
EELV	See Evolved Expendable Launch Vehicle.
EI	See Entry Interface.
Electromagnetic Spectrum	A gathering of all wave-like radiation known to humans. Includes visible light, radio waves, microwaves, gamma rays, and others.
Endeavour	Shuttle orbiter built out of spare parts to replace *Challenger*. First flew in 1992.
Entry Interface	Somewhat arbitrary point set as the start of atmospheric entry for a craft returning from space. Used as reference point for all events in the *Columbia* breakup.
Eros (1443)	An irregularly shaped near-Earth asteroid visited by the Near Earth Asteroid Rendezvous (NEAR) mission from 2001-02.
ERV	See Earth Return Vehicle.
ESA	See European Space Agency.
ET	See External Tank.
European Space Agency	The European counterpart to NASA. Made up of contributions from member states, ESA is best known for its Ariane series of launchers, but has done some space research in concert with other nations and on its own.

Evolved Expandable Launch Vehicle	A US Air Force program designed to cut launch costs by up to one half. Meant to take advantage of rising commercial space activity, the rise in activity did not take place leading to a greater cost per flight than originally hoped for.
External Tank	A portion of the space shuttle stack. The External Tank holds the propellants (liquid hydrogen and liquid oxygen) that the space shuttle main engines burn en route to orbit.
Foreign Object Debris	A standard aeronautical term for any object that could strike an aircraft. The CAIB recommended that NASA use this term for all debris that could strike the space shuttle.
Galactic Cosmic Rays	High energy particles that approach from all directions. Shielding a spacecraft in flight against GCR is not practical with today's technology, and the long-terms effects it has on humans are unknown.
Gas Chromatograph	Experiment that flew aboard the *Viking* spacecraft, designed to describe the elements that make up the soil. Found no carbon, which is believed to be required for life.
GCMS	See Gas Chromatograph.
GCR	See Galactic Cosmic Rays.
Gemini	A two-man spacecraft flown by NASA from 1965-66. Designed to prove the concepts that would be useful for Apollo moon missions.
HST	See Hubble Space Telescope.

Hubble Space Telescope	First of NASA's "Great Observatories" the Hubble was launched by the space shuttle in 1990. Although a flawed mirror delayed the realization of its eventual capabilities, a series of repair and upgrade missions kept the telescope functioning.
ICAMSR	See International Committee Against Mars Sample Return.
ICBM	See Intercontinental Ballistic Missile.
Ida	A main belt asteroid. Visited by the *Galileo* probe in the '90s in a flyby, and found to have its own small moon, named Dactyl, Ida is the first asteroid with a proven moon.
Intercontinental Ballistic Missile	A rocket that can deliver a payload (usually an atomic weapon, although other payloads would be possible) anywhere on the planet. The flight time can be as short as 30 minutes.
International Committee Against Mars Sample Return	Group of scientists and other concerned individuals against the idea of bringing parts of Mars back to Earth without very stringent biological safeguards in place.
International Space Station	The largest project ever attempted in space in terms of mass and expenditure. International partners including the United States, Russia, Europe, Japan, Canada, and Brazil are constructing the station, designed as an orbital outpost for experimenting on zero-gravity's effect on humans, as well as early industrial processes.
Isp	See Specific Impulse.

ISS	See International Space Station.
JIMO	See Jupiter Icy Moons Orbiter.
Jupiter Icy Moons Orbiter	A nuclear-powered spacecraft designed to do an in-depth analysis of Jupiter and its moons. Advanced propulsion will allow the craft to orbit each moon and use on-board radar to provide information about water content.
Kennedy Space Center	The NASA center that launches the space shuttle. Located on the East Coast of Florida, it borders Cape Canaveral Air Force Station.
KSC	See Kennedy Space Center.
Labeled Release	A test that flew on board the *Viking* spacecraft, designed to determine if there were bacteria in Martian soil. Returned results originally thought to indicate life, but later results from the gas chromatograph contradicted those results.
LEO	See Low Earth Orbit.
Lockheed Martin	Major military and space contractor. Known for building deep space probes, as well as launch vehicles such as Atlas and Titan, and the Atlas V, a version of the Evolved Expendable Launch Vehicle.
Low Earth Orbit	Though there's no formal definition, low Earth orbit is widely accepted as an orbit around the Earth that is stable, and most do not enter the Van Allen radiation belts.
LR	See Labeled Release.

Lunar Prospector	A small spacecraft that orbited the moon mapping minerals and water on the lunar surface. Showed strong indications of water ice in permanently shadowed areas of the lunar south pole.
Mars Climate Orbiter	Spacecraft sent to Mars in 1999, designed to orbit the planet and monitor Mars' short-term weather and long-term climactic changes. A mix-up between metric and English units led to the spacecraft approaching Mars too closely, it entered Mars' atmosphere and likely crashed on the surface.
Mars Direct	Mars mission architecture proposed by Zubrin and Baker. Involves separating crew return from Mars and crew travel to Mars into two launches. In-situ resource utilization allows a Mars mission to take place using smaller launch vehicles and less on-orbit mass than other architectures.
Mars Exploration Rover	Mars mission involving two identical rovers sent to opposite sides of the planet. The goal of the mission was to find evidence that portions of Mars' surface formed in the presence of liquid water.
Mars Global Surveyor	Spacecraft that arrived in Mars orbit in 1997. Still functioning, this craft has photographed the entire surface at low resolution and continues to take more images at high resolution. Also used to relay information from other surface craft back to Earth.
Mars Odyssey	A mission to Mars launched in 2001. Designed to map possible underground water on Mars, as well as other mineral concentrations.

Mars Pathfinder Flagship of NASA's faster-better-cheaper approach to building spacecraft, the *Mars Pathfinder* was mainly designed to prove that Mars missions could be conducted for much less money than before. Returned amazing surface images, and carried a small rover named Sojuourner.

Mars Polar Lander Mission sent to Mars in 1999. Designed to land on Mars pole and examine the ice there. Crash landed on the surface and sent no meaningful communications back.

Mars Reconnaissance Orbiter Mars mission slated for flight in 2005. Scheduled to carry a more powerful camera into Mars orbit than ever before, allowing pictures of unprecedented detail.

Mars Scout Mission A series of missions suggested by universities and research organizations outside NASA. First is scheduled to fly in 2007, with others expected to follow.

MCO See *Mars Climate Orbiter*.

MER-A (*Spirit*) The first Mars Exploration Rover launched. Landed in Gusev Crater on Mars on January 4th, 2004.

MER-B (*Opportunity*) The second Mars Exploration Rover, launched in July and landing on Meridiani Planum on January 24th (ET), 2004.

Mercury (capsule) A one-man spacecraft flown by NASA from 1962-63.

MGS See *Mars Global Surveyor*.

MPF See *Mars Pathfinder*.

MPL See Mars Polar Lander.

MRO	See Mars Reconnaissance Orbiter.
NACA	See National Advisory Council on Aeronautics.
NASA Select	Television channel run by NASA showing live images of shuttle missions and other tests, big press conferences, and historical footage as filler.
NASP	See National Aerospace Plane.
National Advisory Council on Aeronautics	Precursor to NASA. NACA was founded in 1915 and set research standards along with funding development of critical devices and practices in the field of aviation.
National Aerospace Plane	An effort, started during the Reagan administration, to make an aircraft that can fly to the edges of the atmosphere at very high speed. After some early development work, the program was scrapped.
National Imagery and Mapping Agency	Now known as the National Geospatial-Intelligence Agency. US Government organization involved in gathering information from national assets with classified capabilities.
National Launch System	An advanced launcher system that was a collaborative effort between NASA and the Air Force. Took several different forms in redesign, and was eventually canceled.

National Oceanographic and Atmospheric Administration	US Government organization tasked with monitoring US weather among other duties. Controls a fleet of satellites to monitor weather on Earth and in space.
NEAR	See Near Earth Asteroid Rendezvous.
Near Earth Asteroid Rendezvous	One of NASA's Discovery class missions, this craft flew to the asteroid Eros, orbiting the space rock for over a year and then landing on the surface.
Near Earth Object	A classification of space rocks that include asteroids and comets that cross Earth's orbit.
NEO	See near Earth object.
NIMA	See National Imagery and Mapping Agency.
NLS	See National Launch System.
NOAA	See National Oceanographic and Atmospheric Administration.
NTR	See Nuclear Thermal Rocket.
Nuclear Thermal Rocket	Although billed as advanced propulsion, working versions of the NTR were actually developed in the 1960s. Design passes a fluid (usually hydrogen) over a nuclear reactor and ejects the byproducts. Produces more thrust per unit mass of fuel than chemical propulsion, and allows larger payloads to fly.
Office of Management and Budget	Division of the presidential staff that examines and prepares budget documentation and analysis.

OMB	See Office of Management and Budget.
Orbital Space Plane	A NASA program meant to provide an escape capsule for ISS crews, as well as a new route into space for personnel after retirement of the Space Shuttle. The Bush space announcement of January 2004 cancelled the OSP in favor of the Crew Exploration Vehicle.
OSP	See Orbital Space Plane.
Phobos	Moon of Mars.
Phoenix	The first Mars Scout Mission selected. The project re-uses the lander from the cancelled 2001 mission, and will land near the Martian north pole. Research plans include investigation of the environment under the polar ice.
Processing Debris	A term NASA coined for debris that struck the space shuttle but only caused extra preparation requirements. The CAIB stated that such terms made debris strikes seem like common events when each one should have been taken very seriously.
Progress	A Russian uncrewed cargo vehicle, based on the *Soyuz* design. Used to transfer supplies to Russian space stations in the past, and the International Space Station today.
Radioisotope Thermoelectric Generator	Power source used to power deep-space probes beyond the asteroid belt. Uses the heat energy produced in the decay of plutonium to heat the craft and generate electricity
RCC	See reinforced carbon-carbon.
Regolith	Scientific term for dirt on the moon and on Mars.

Reinforced Carbon-Carbon	Extremely heat-resistant and seemingly tough material placed at spots that receive the most heat during a space shuttle reentry. RCC at the leading edge of *Columbia*'s wing was damaged on launch, leading to the orbiter's breakup in February 2003.
Reverse-Water Gas Shift	A chemical reaction that combines carbon dioxide and water in a repeated process to produce a constant stream of oxygen, suitable for breathing. If no hydrogen leakage is assumed, a RWGS reactor can produce oxygen as long as power is applied.
RTG	See Radioisotope Thermoelectric Generator.
RWGS	See Reverse-Water Gas Shift.
Sabatier Reaction	Chemical process that combines carbon dioxide and hydrogen in the presence of a catalyst to produce methane (a rocket fuel) and water. Proposed as part of a reduced-cost crewed Mars mission.
SCWO	See Super-Critical Water Oxidation.
Skylab	An orbital workshop launched into Earth orbit in a single flight in 1973. Visited by three crews of three men each for stays of 28, 56, and 84 days. Fell to Earth in 1979.
SLI	See Space Launch Initiative.
Sojourner	Small rover carried aboard *Mars Pathfinder*. Proved that rovers could be built small and provided the first images of a US uncrewed spacecraft.
Solar Flare	An explosion on the surface of the sun that blasts vast amounts of gas and radiation into space. Usually related to a coronal mass ejection.

Solar Maximum	Peak of the 11-year solar cycle. Solar activity including flares, sunspots, and coronal mass ejections is at its highest. In some ways, this would be the most dangerous time to launch a space mission.
Solar Minimum	Low-point of the 11-year solar cycle. Solar activity is minimized. Leads to increased exposure to galactic cosmic rays for a space mission.
Solid Rocket Booster	A portion of the space shuttle stack. Two Solid Rocket Boosters provide the primary thrust to lift the stack from the surface of the Earth. After firing, they fall away into the ocean and after a lengthy refurbishment are reused.
Soyuz	A Russian crewed spacecraft. First flown in 1967, the spacecraft has been upgraded several times, and currently serves primarily as a ferry and rescue craft for the International Space Station.
Space Launch Initiative	A NASA program designed to carry out research and design of the next generation of space launch vehicle. Consisted of a number of study contracts let to different vendors. Eventually, this effort was cancelled, so that NASA could focus on developing the Orbital Space Plane.
Space Transportation System	The formal name for the space shuttle. The term is abbreviated as STS, then used to preface all mission numbers.
SpaceShipOne	Rocket-powered craft that achieves high altitude as part of Scaled Composites' X-prize entry. Powered by rubber and laughing gas, it can carry 3 people.
Specific Impulse	The measure of a rocket fuel's efficiency. Defined as the amount of fuel it takes for a type of fuel to maintain one Newton of thrust for one second.

SRB	See Solid Rocket Booster.
STS	See Space Transportation System.
Super-critical water oxidation	A waste reduction method that uses water under intense heat and pressure to break waste material down into its component atoms for ease of re-use or disposal.
Ultra-Violet	A portion of the electromagnetic spectrum. More energetic than visible light, ultra-violet rays cause sunburns and can be used to remove bacteria from materials or liquids.
Union of Soviet Socialist Republics	Also less correctly known as Russia, was the primary adversary to the United States during the cold war. Their early space successes spurred the US into action.
United Space Alliance	Company formed as the combined effort of Lockheed Martin and Boeing to operate the space shuttle. Part of a NASA effort to cut the costs of operating the space shuttle.
USA	See United Space Alliance.
USSR	See Union of Soviet Socialist Republics.
UV	See ultra-violet.
Van Allen radiation belts	Zones of charged particles trapped by Earth's magnetic field. Discovered by the United States' first satellite, Explorer 1.

Venturestar	The larger-scale version of the X-33. Original plans had the contractor (Lockheed Martin) building Venturestar after the X-33. This craft was never built.
White Knight	Carrier craft for the Scaled Composites' entry in the X-prize competition. Drops the *SpaceShipOne* rocket craft.
X Prize	Founded by Peter Diamandis, the X-prize is based on aeronautical prizes of decades past. Pays $10M to first privately developed craft to carry three people to 100 km, then repeat the feat within 14 days.
X-33	A spacecraft demonstrator designed to test the theory of single-stage to orbit. The X-33 would have launched like a rocket and flown to high altitudes and speed, then returned to Earth and landed like an aircraft. The X-33 was cancelled after budget overruns and some test failures in critical components.
X-34	A spacecraft demonstrator, designed to be dropped from an aircraft, fly to high altitude and speed, and then return to Earth. Cancelled in a round of budget cuts.

INDEX

Aldridge, Pete, 37, 259
Allen, Paul, 55
Alternate Access to Space, 153, 166
American Institute of Aeronautics and Astronautics, 132, 208
Apollo, 13, 3, 5, 30, 31, 33, 34, 40, 41, 43, 63, 78, 81, 96, 102, 110, 112, 116, 127, 131, 162, 169, 172, 174, 175, 183, 185, 186, 188, 190, 208, 212, 217, 220, 221, 225, 227, 228, 245-247, 256, 259, 263
Armadillo Aerospace, 55
Artemis Society, 132, 139, 140
Arthur C. Clarke, 20, 25, 32, 131, 154, 193
Asteroid, 7, 45, 80, 96, 98, 136, 168-170, 179, 180, 182, 184, 185, 190, 192-201, 211, 219, 224, 226, 233, 247, 248, 250, 254, 255
Automated Transfer Vehicle, 11, 90, 175, 252
B612 Foundation, 136
Bezos, Jeff, 21, 235
Blue Origin, 21, 55, 235
Boeing, 52, 59, 60, 84, 150, 270
Buran, 252
Cape Canaveral Air Force Station, 2, 272
Carmack, John, 55

Cassini, 4, 12, 123, 142-144, 159, 275, 276, 279, 283
Centre National d'Etudes Spatiales, 10
Challenger, 37-39, 41, 42, 50, 54, 63, 78, 93, 96, 100, 103, 112, 131, 165, 252
China, 13, 14, 47, 226, 272
Columbia, 11, 13, 4, 5, 9, 33, 34, 54, 61-72, 81-92, 96, 103, 105, 110, 112, 126, 144, 145, 147, 148, 151, 164-166, 229, 251, 252, 257, 261, 269
Columbia Accident Investigation Board, 11, 5, 8, 61, 78, 85, 89, 92, 165, 223, 229, 251, 257, 261
COMSAT, 33, 231, 238, 239
Corona, 27-29, 113
Coronal Mass Ejection, 173, 174
Crew Exploration Vehicle, 3, 111, 112, 162, 166, 258
Devon Island, 22, 23, 137, 140
Discovery, 42, 108, 260
Entry Interface, 64-66, 72
European Space Agency, 10, 35
Evolved Expendable Launch Vehicle, 51, 111, 155, 162, 166, 230
Falcon, 18, 19, 117
Gehman, Harold, 67, 92, 252
Gemini, 30, 31, 40, 112, 137, 162, 173, 246

Global Positioning System, 1, 12, 60, 113, 121
High Earth Orbit, 168-171, 175, 177-179, 184, 199, 214, 248, 254, 255
Hubble Space Telescope, 4-6, 92, 187, 269
India, 14, 193, 197
Institute of Space and Astronautical Science, 12
International Committee Against Mars Sample Return, 144, 216
International Space Station, 1, 4, 8, 13, 43, 44, 53, 54, 63, 67, 81, 128, 147, 150, 151, 165, 166, 175, 223, 252, 253, 263, 269
Iridium, 44, 47-49, 113, 115, 118
Japanese Aerospace Exploration Agency, 13
Japanese Experimental Module, 13
JIMO, 123, 279
Kennedy Space Center, 2, 23, 39, 67, 93, 253, 272
Launch Window, 255, 273-276, 279
Low Earth Orbit, 1 2, 8, 14, 18, 19, 24, 36, 61, 105, 115, 154, 155, 158, 159, 164, 169, 172, 177, 191, 201, 217, 221, 229, 230, 235, 236, 246, 248-251, 253-255, 258, 268, 270, 277
Lowell, Percival, 202
Mars Direct, 43, 137, 159, 207, 208

Mars Exploration Rover, 56, 72, 154, 206, 210, 260, 261, 278
Mars Global Surveyor, 46, 56, 79, 204, 210, 278
Mars Institute, 139, 140, 145
Mars Society, 22, 23, 132, 137-140, 145, 177, 207, 253
Mercury, 30, 40, 104, 112, 192, 273
Mir, 8, 44, 78, 153, 165, 175, 176
Modular Auxiliary Data System, 64, 66-68, 70, 71
Moon Society, 139, 140
Musk, Elon, 18, 138, 139, 235
National Advisory Council on Aeronautics, 7, 242
National Aeronautics and Space Administration, 1, 3, 6, 16, 39, 66, 67, 69, 72, 91, 111, 116, 137, 138, 140, 142-144, 165, 166, 208, 227-230, 234, 236, 238, 242, 243, 245, 246, 248-250, 253, 254, 257-259, 261, 262, 269, 270, 272, 275, 278, 280
National Aerospace Laboratory, 12
National Aerospace Plane, 241
National Launch System, 241
National Oceanographic and Atmospheric Administration, 1, 13, 248
National Space Club, 137
National Space Development Agency, 12

301

Near Earth Object, 192, 195, 201, 226, 247
OKeefe, Sean, 66, 259
Orbital Space Plane, 166, 258
Orbital Supply Depots, 248
Park, Robert, 127
Plait, Philip, 175
Planetary Society, 24
Progress, 141, 252
ProSpace, 141
Radioisotope Thermoelectric Generator, 279
Reinforced Carbon-Carbon, 69, 71
Reusability, 270
Robert Heinlein, 221, 253
Rutan, Burt, 55, 241
Simberg, Rand, 150
Skylab, 13, 34, 36, 151, 165, 175
Slingshots, 274
Solar sail, 24
Solid Rocket Booster, 264
Soyuz, 9, 10, 14, 34, 54, 78, 90, 96, 104, 112, 147, 164, 251, 252
Space Transportation System, 36, 37, 74, 88, 93, 109, 165, 248
SpaceShipOne, 15, 55, 241, 282
SpaceX, 18, 117, 138, 222, 235
Specific Impulse, 265, 266, 277
The Aerospace Corporation, 259
Ulysses, 275
Viking, 35, 277
Von Braun, Werner, 246
White Knight, 55
X-33, 241
X-34, 241
X-Prize, 15, 49, 55, 222, 282, 283

Zubrin, Robert, 127, 137, 138, 186, 224, 253, 263, 276

REFERENCES

[1] http://www.space.com/news/nasa_budget_040130.html accessed February 9, 2004

[2] http://a255.g.akamaitech.net/7/255/2422/02feb20041242/www.gpoaccess.gov/usbudget/fy05/pdf/budget/tables.pdf accessed February 9, 2004

[3] Thomas, Cathy Booth, et al. "Mission to Mars," *Time Magazine*, January 26, 2004.

[4] Budget of the US Government, FY 2005: Budget documents http://www.whitehouse.gov/omb/budget/fy2005/budget.html accessed February 24, 2004

[5] http://spaceflightnow.com/stardust/031230return.html accessed February 9, 2004

[6] http://www.savethehubble.org accessed February 21, 2004

[7] Statistical Abstract of the United States, 2002, Chart 778.

[8] http://www.house.gov/science_democrats/member/bg030611.htm accessed February 9, 2004

[9] http://www.fas.org/nuke/guide/russia/c3i/ accessed February 9, 2004

[10] Two Americans Sign up for Space Trips http://msnbc.msn.com/id/3732412/ accessed February 9, 2004

[11] Arianespace and the Ariane Family of Rockets http://www.centennialofflight.gov/essay/SPACEFLIGHT/ariane/SP42.htm accessed February 14, 2004

[12] Ariane V G http://www.centennialofflight.gov/essay/SPACEFLIGHT/ariane/SP42.htm and Ariane 5 Failure Focuses on Upper Stage RLINK

"http://www.spaceflightnow.com/ariane/v142/010713followup.html" http://www.spaceflightnow.com/ariane/v142/010713followup.html accessed February 14, 2004

[13] *Mars Express* sees its first water http://www.esa.int/export/esaCP/SEM8ZB474OD_index_0.html accessed February 14, 2004

[14] Three Space Agencies to be Integrated http://www.japantoday.com/gidx/news56502.html accessed February 14, 2004

[15] Economy slows Japan's space program http://www.abc.net.au/news/newsitems/s862909.htm accessed February 14, 2004

[16] The X-Prize Website http://www.xprize.org/ accessed February 14, 2004

[17] The Cheap Access to Space Prize http://www.space-frontier.org/Events/CATSPRIZE_1/ accessed February 14, 2004

[18] Say no to NASA, yes to private companies http://www.usatoday.com/usatonline/20030924/5528189s.htm accessed February 14, 2004

[19] DARPA's Grand Challenge http://www.darpa.mil/grandchallenge/ accessed February 21, 2004

[20] Falcon Overview http://spacex.com/ accessed February 14, 2004

[21] Space Launch Vehicle Reliability http://www.aero.org/publications/crosslink/winter2001/03.html accessed February 14, 2004

[22] http://www.teamencounter.com accessed February 21, 2004

[23] Clarke, Arthur C. *The Wind from the Sun; Stories of the Space Age*, copyright 1972, Harcourt Brace Jovanovich.

[24] New ways to fund space ventures http://thespacereview.com/article/60/1 accessed February 14, 2004

[25] http://www.bigelowaerospace.com

[26] Sheet Metal Workers Become Mars Desert Research Station Sponsors http://www.marssociety.org/bulletins/06.18.01.Bulletin44.2.asp accessed February 14, 2004

[27] http://www.marssociety.org accessed February 24, 2004

[28] Solar Sail Update: Launch Planned for Early 2004 http://www.planetary.org/solarsail/update_20030919.html accessed February 14, 2004

[29] Ruffner, Kevin C, Editor *CORONA: America's First Satellite Program*, CIA Cold War Records, available through the Photoduplication Service, Library of Congress, Washington, DC, 20540.

[30] Air Force Underground http://www.af.mil/news/airman/0897/oscar2.htm, accessed February 14, 2004

[31] Belland, L and Wells R *Space Satellite: The Story of the Man-Made Moon*, Prentice-Hall, Inc, 1957.

[32] Hickam, Homer. *Rocket Boys*, Delacorte Press 1998.

[33] A Brief History of Amateur Satellites http://amsat.org/amsat/sats/n7hpr/history.html, accessed February 14, 2004

[34] Baker, David. *The History of Manned Spaceflight*, New Cavendish Books, 1985 pp 243.

[35] Clark, Phillip. *The Soviet Manned Space Program*, Salamander Books Limited, 1988 pp 64.

[36] Scientists Say Mars Viking Mission Found Life http://www.space.com/news/spacehistory/viking_life_010728-1.html accessed February 15, 2004

[37] Jenkins, Dennis R. *Space Shuttle: The History of Developing the National Space Transportation System, The Beginning through STS-75*, Dennis R. Jenkins, Publisher 2000 pp 125.

[38] Baker, David. *Space Shuttle*, Crown Publishers Inc 1979, pp 63.

[39] US Air Force Lowers Boom on Delta Program http://www.space.com/missionlaunches/boeing_eelv_030724.html accessed February 16, 2004

[40] NASA's Return to Flight Page http://www.nasa.gov/news/highlights/returntoflight.html accessed February 21, 2004

[41] Transterrestrial Musings, Safe Enough? http://www.interglobal.org/weblog/archives/003470.html accessed February 26, 2004

[42] Burrough, p 21-27.

[43] QuickFactFinder, US Census Bureau, Education by sex http://factfinder.census.gov/servlet/QTTable?_bm=y&-geo_id=01000US&-qr_name=DEC_2000_SF3_U_QTP20&-ds_name=DEC_2000_SF3_U&-_lang=en&-_sse=on accessed February 16, 2004

[44] Astronaut Biographies Home Page http://www.jsc.nasa.gov/Bios/ accessed February 16, 2004

[45] http://www.nasa.gov/home/hqnews/2003/jun/HQ_03203_Edu.Ast_Apl.html accessed February 21, 2004

[46] Fly me to the Moon and On to Mars http://www.foxnews.com/story/0,2933,111158,00.html accessed February 16, 2004

[47] http://www.jsc.nasa.gov/bios/htmlbios/garn-j.html accessed February 21, 2004

[48] http://www.jsc.nasa.gov/Bios/htmlbios/nelson-b.html accessed February 21, 2004

[49] http://www.jsc.nasa.gov/Bios/htmlbios/al-saud.html accessed February 21, 2004

[50] http://www.interglobal.org/weblog

[51] Pellegrino, C.R. and Stoff, J. *Chariots for Apollo: The Untold Story Behind the Race to the Moon*, Avon Books Inc. 1985 pp 280-283.

[52] http://world.honda.com/ASIMO/

[53] Robert Zubrin, in debate with Robert Park on the merits of human spaceflight. Transcript at http://eppc.org/conferences/pubID.2029,eventID.70/transcript.asp

[54] http://www.b612foundation.org accessed February 21, 2004

[55] http://www.astronautix.com/flights/gemini8.htm accessed February 16, 2004

[56] Mars Gravity Biosatellite Program http://www.marsgravity.org/main/ accessed February 16, 2004

[57] Profile of Elon Musk http://www.space-frontier.org/Projects/Spacefaring/1-9%20Elon%20Musk.htm accessed February 16, 2004

[58] http://www.moonsociety.org

[59] http://www.asi.org

[60] Simberg, Rand Transterrestrial Musings – Into the Wilderness http://www.foxnews.com/story/0,2933,77783,00.html accessed February 16, 2004

[61] Smitherman, et al Space Resource Requirements for Future In-space Propellant Production Depots http://www.mines.edu/research/srr/2001Program.htm accessed February 21, 2004

[62] Space Development: The Case Against Mars http://www.foresight.org/NanoRev/Mars.html accessed February 16, 2004

[63] Plait, Philip. *Bad Astronomy: Misconceptions and Misuses Revealed, from Astrology to the Moon Landing "Hoax"*. John Wiley & Sons, 2002. Also http://badastronomy.com/bad/tv/foxapollo.html accessed February 16, 2004 for moon-landing hoax information

[64] Stuster, Jack. *Bold Endeavors: Lessons from Polar and Space Exploration*, The Navy Press, 1996.

[65] Weed, William Speed. "The Worst Jobs in Science," *Popular Science*, October 2003.

[66] 2001: A Space Odyssey Revisited – The Feasibility of 24 Hour Commuter Flights… http://astp.msfc.nasa.gov/ast/abstracts/7A_Barowski.html Accessed February 27, 2004

[67] Zubrin, Robert. *Entering Space: Creating a Spacefaring Civilization*, pp 87 Tarcher/Putnam 1999.

[68] ibid, pp 89-90.

[69] Eckart, Peter. *Spaceflight Life Support and Biospherics* pp 241, Space Technology Library 1996.

[70] Christianity: Came from Outer Space http://dsc.discovery.com/news/briefs/20030623/constantine.html accessed February 16, 2004

[71] Near-Earth Object Threat, testimony given by Simon P. Worden to the House Science Committee, October, 2002 http://www.house.gov/science/hearings/space02/oct03/worden.htm

[72] Report of the Task Force on potentially hazardous Near Earth Objects http://www.nearearthobjects.co.uk/report/resources_task_intro.cfm accessed February 19, 2004

[73] Hitting the Sweet Spot
http://www.wired.com/wired/archive/11.11/newsugar_pr.html accessed February 19, 2004

[74] Zubrin Robert. *The Case for Mars: The Plan to Settle the Red Planet and why we Must*, The Free Press, 1996 pp153.

[75] Donahue B, and Cupples M. "Comparative Analysis of Current NASA Human Mars Mission Architectures," *AIAA Journal of Spacecraft and Rockets*, Sep-Oct 2001 pp 745.

[76] Zubrin, *Case for Mars* pp 182.

[77] The New Red Menace
http://www.wired.com/wired/archive/9.07/mars.html?pg=3&topic=&topic set= accessed February 19, 2004

[78] Savings and Loan Crisis
http://www.econlib.org/library/Enc/SavingsandLoanCrisis.html accessed February 19, 2004

[79] Treaty on Principles Governing the Activities of States in the Exploration and Use of Outer Space, Including the Moon and Other Celestial Bodies
http://www.state.gov/t/ac/trt/5181.htm

[80] THOMAS – U.S. Congress on the Internet
http://thomas.loc.gov/, search for HR 3752. accessed February 21, 2004

[81] Bill Gates Net Worth Page
http://www.quuxuum.org/~evan/bgnw.html accessed February 21, 2004

[82] Sweetman, Bill. "The Contender: How Airbus got to be Number one," *Air&Space Smithsonian Magazine*, Oct/Nov 2003.

[83] J-Track 3D
http://liftoff.msfc.nasa.gov/RealTime/JTrack/3D/JTrack3D.html accessed February 21, 2004

[84] By Gemini to the Moon!
http://www.astronautix.com/articles/bygemoon.htm accessed February 21, 2004

[85] Russian Soyuz One Step Closer to Launch from French Guiana
http://www.space.com/missionlaunches/kourou_soyuz_031107.html accessed February 21, 2004

[86] NASA reclaims parts from museum exhibit
http://www.cnn.com/TECH/space/9902/15/nasa.museum/ accessed February 21, 2004

[87] President Bush Announces New Vision for Space Exploration Program
http://www.whitehouse.gov/news/releases/2004/01/20040114-3.html accessed February 21, 2004

[88]
http://boeingmedia.com/images/search.cfm?product_id=1525 accessed February 21, 2004

[89] http://www.moontomars.org/ accessed February 21, 2004

[90] Special Message to Congress on Urgent National Needs
http://www.cs.umb.edu/jfklibrary/j052561.htm accessed February 22, 2004

[91] Zubrin, Robert. *Entering Space*, pp 168-170.

Notes

Printed in the United States
65603LVS00005B/139